Arthur S. Eddi

Report on the Relativity Theory of Gravitation

With a Foreword by Paul S. Wesson

MINKOWSKI
Institute Press

Arthur S. Eddington
28 December 1882 – 22 November 1944

Cover: One of the photographs of the total solar eclipse of 29 May 1919, discussed in F. W. Dyson, A. S. Eddington and C. Davidson, "A Determination of the Deflection of Light by the Sun's Gravitational Field, from Observations Made at the Total Eclipse of May 29, 1919," *Phil. Trans. R. Soc. Lond. A* 1920 **220**, pp. 291-333.
Photograph taken from: `http://en.wikipedia.org/wiki/Arthur_Eddington`

ISBN: 978-1-927763-28-5 (softcover)
ISBN: 978-1-927763-29-2 (ebook)

Minkowski Institute Press
Montreal, Quebec, Canada
http://minkowskiinstitute.org/mip/

For information on all Minkowski Institute Press publications visit our website at http://minkowskiinstitute.org/mip/books/

Arthur Stanley Eddington in 1931
(Source: http://silas.psfc.mit.edu/eddington/)

Preface to the 2014 Publication

This volume contains new publications of two important works:

- A. S. Eddington, *Report on the Relativity Theory of Gravitation*, 2ed (Fleetway Press, London 1920).
- F. W. Dyson, A. S. Eddington and C. Davidson, A Determination of the Deflection of Light by the Sun's Gravitational Field, from Observations Made at the Total Eclipse of May 29, 1919, *Phil. Trans. R. Soc. Lond.* A 1920 **220**, pp. 291-333.

These works have never been published together and are two firsts: Eddington's *Report*, whose first edition appeared in 1918, was the first systematic exposition of Einstein's general relativity not only in English. The article by Dyson, Eddington, and Davidson reported on the first experimental verification of general relativity.

There were some doubts in 1980 on whether the data of this first experimental test of general relativity were properly and unbiasedly analyzed. But it turned out that in 1978 an reanalysis of the eclipse data reported by Dyson, Eddington, and Davidson had already confirmed that the deflection of light by the Sun's gravitational field had been accurately observed (see [1-2] and the references therein).

It may appear natural to ask "Why should a book on general relativity published two years after the publication of general relativity itself be reprinted again given that there exist hundreds of excellent books on it?" The answer to this question was given in 1983 by the renowned astrophysicist Subrahmanyan Chandrasekhar who received the 1983 Nobel Prize for Physics "for his theoretical studies of the physical processes of importance to the structure and evolution of the stars" [3]:

> It will be recalled that in the last of the communications in which Einstein formulated his fundamental field equations, he concluded with the prophetic statement, 'scarcely anyone who has fully understood this theory can escape from its magic.' Eddington must surely have been caught in its magic; for, within two years, he had written his *Report on the Relativity Theory of Gravitation* for the Physical Society of London, a report that must have been written in white heat. Eddington's *Report* is written so clearly and yet so concisely that it can be read, even today, as a good introductory text by a beginning student.

Montreal, 5 November 2014 *Vesselin Petkov*

1. D. Kennefick, "Testing relativity from the 1919 eclipse – a question of bias," *Physics Today*, March 2009, pp. 37–42

2. D. Kennefick, "Not Only Because of Theory: Dyson, Eddington and the Competing Myths of the 1919 Eclipse Expedition," in: C. Lehner, J. Renn, and M. Schemmel (eds), *Einstein and the Changing Worldviews of Physics* (Springer, Heidelberg 2012), pp. 201-232

3. S. Chandrasekhar, *Eddington – The most distinguished astrophysicist of his time* (Cambridge University Press, Cambridge 1983), p. 24

Foreword

Eddington published *Report on the Relativity Theory of Gravitation* in 1918. His purpose was in part to make Einstein's recently formulated general theory of relativity available to the English-speaking world. In this, he certainly succeeded. The book covers the classical and therefore timeless aspects of the theory in Eddington's typically eloquent style. (Naturally, it is silent about the flood of exact data for the universe that we have experienced in the last decade or so.) Only a couple of years after Eddington's book came out, Einstein himself published a small volume in English called *The Meaning of Relativity*. However, the two books should not be seen as in competition with each other. The one is a technical account aimed at the physicist, while the other is meant for a wider audience. It is a testament to both men that their books are still sought after.

Eddington had a profound understanding of Einstein's theory of general relativity, despite its reputation for difficulty. (One reporter asked Eddington if it were true that the theory was only understood by three people, to which the scientist responded by asking who the third person might be.) However, while expert in the most esoteric forms of theoretical physics, he also had hands-on experience with data. For example, he travelled to the tropics to help with observations of a total eclipse, one of two expeditions designed to test the new theory. This was in May 1919, a year after his longer *Report* first appeared. The accuracy of the observations was not the best, but was good enough to confirm the prediction that light rays would be bent as they passed near the Sun. The results of this trip were published by the Royal Society in 1920 and the article by Dyson, Eddington and Davidson is reprinted in the present volume. At his desk in Cambridge, Eddington was thus able to devote himself to working out the consequences of general relativity, secure in the knowledge that the theory worked. At that time he was Plumian Professor at Cambridge University, and lived in the imposing structure which is now mainly used as the library for the Institute of Astronomy. (When the writer was a graduate student, afternoon tea was still prepared in what had been Eddington's kitchen, no doubt supervised by his sister with whom he lived.) His skill at mathematics, combined with an uncanny insight to the workings of Nature, caused Eddington to make a series of major contributions to the structure of the theory, as well as its applications to astrophysics and cosmology. Along with Einstein, he was sure that the general theory of relativity gave a superior account of the world, especially insofar as gravity was concerned.

However, while the two great men agreed on many things, they disagreed about the cosmological constant. This parameter was initially introduced by Einstein to account for a static universe, but later abandoned by him when the galaxies were found to be fleeing from each other. He continued to denigrate the parameter, calling it the biggest blunder of his life. By contrast, Eddington was totally accepting of it. He went on to write in his popular book *The Expanding*

Universe that to drop the cosmological constant would "knock the bottom out of space." By this Eddington meant that the parameter was not only vital in cosmology, but also played a role in certain more speculative accounts of quantum mechanics with which he was concerned in the later part of his life.

It is nowadays often the case that Eddington's later work, preoccupied as it was with numerology, is politely dismissed as too fanciful to be taken seriously. But as one biographer has commented, Eddington was probably aware that he was headed to an early grave; and his later researches should be viewed as a rushed attempt to describe his scientific beliefs in the shortest possible way. It is also worth pointing out that according to the best modern data, the universe *is* dominated by the cosmological constant. Since Eddington has been shown correct in his views about that, it is conceivable that he may prove to be right about some other things …

For now, the reader may rest assured that Eddington's *Report* is an accurate account of general relativity, and that his old observations have been augmented by a wealth of new data which show the theory to be in good agreement with the real universe.

Paul S. Wesson
Professor, University of Waterloo

Preface to First Edition

The relativity theory of gravitation in its complete form was published by Einstein in November 1915. Whether the theory ultimately proves to be correct or not, it claims attention as one of the most beautiful examples of the power of general mathematical reasoning. The nearest parallel to it is found in the applications of the second law of thermodynamics, in which remarkable conclusions are deduced from a single principle without any inquiry into the mechanism of the phenomena; similarly, if the principle of equivalence is accepted, it is possible to stride over the difficulties due to ignorance of the nature of gravitation and arrive directly at physical results. Einstein's theory has been successful in explaining the celebrated astronomical discordance of the motion of the perihelion of Mercury, without introducing any arbitrary constant; there is no trace of forced agreement about this prediction. It further leads to interesting conclusions with regard to the deflection of light by a gravitational field, and the displacement of spectral lines on the sun, which may be tested by experiment.

The arrangement of this Report is guided by the object of reaching the theory of these crucial phenomena as directly as possible. To make the treatment rather more elementary, use of the principle of least action and Hamiltonian methods has been avoided; and the brief account of these in Chapter VII. is merely added for completeness. Similarly, the equations of electro-dynamics are not used in the main part of the Report. Owing to the historical tradition, there is an undue tendency to connect the principle of relativity with the electrical theory of light and matter, and it seems well to emphasize its independence. The main difficulty of this subject is that it requires a special mathematical calculus, which, though not difficult to understand, needs time and practice to use with facility. In the older theory of relativity the somewhat forbidding vector products and vector operators constantly appear. Happily this can now be avoided altogether; but in its place we use the absolute differential calculus of Ricci and Levi-Civita. This is developed *ab initio* so far as required in Chapter III. Attention must be called to the remark on notation in §19, which concerns almost all the subsequent formulae.

Extensive use has been made of the following Papers, which in some places have been followed rather closely:

A. Einstein. Die Grundlage der allgemeinen Relitivitäts theorie "Annalen der Physik," 57 XLIX., p. 769 (1916).

W. de Sitter. On Einstein's Theory of Gravitation and its Astronomical Consequences. "Monthly Notices of the Royal Astr. Soc.," LXXVI, p. 699 (1916); LXXVII, p, 155 (1916); LXXVIII, p. 3 (1917).

I am especially indebted to Prof. de Sitter, who has kindly read the proof-sheets of this Report.

The principal deviations in the present treatment of the subject will be found in Chapter VI. I have ventured to modify the enunciation of the principle of equivalence in §27 in order to give a precise criterion for the cases in which it is assumed to apply.

Other important Papers on the subject, most of which have been drawn on to some extent, are:

D. HILBERT. Die Grundlagen der Physik, "Göttingen Nachrichten," 1915, Nov. 20.

H. A. LORENTZ. On Einstein's Theory of Gravitation, "Proc. Amsterdam Acad.," XIX., p. 1341 (1917).

J. DRÖSTE. The Field of n moving centres on Einstein's Theory, "Proc. Amsterdam Acad.," XIX., p. 447 (1916).

A. EINSTEIN. Kosmologische Betrachtungen zur allgemeinen Relitivitätstheorie, "Beilin Sitzungsber.," 1917, Feb. 8. Ueber Gravitationswellen, *ibid*, 1918, Feb. 14.

K. SCHWARZSCHILD. Ueber das Gravitationsfeld eines Massenpunktes nach der Einstein'schen Theorie, "Berlin Sitzungsber," 1916, Feb. 3.

T. LEVI-CIVITA. Statica Einsteiniana, "Rendiconti dei Lincei," 1917, p, 458.

A. PALATINI. Lo Spostamento del Perielio di Mercurio, "Nuovo Cimento," 1917, July.

The last two Papers avoid much of the heavy algebra, but claim a rather extensive knowledge of differential geometry.

The older theory of relativity, briefly surveyed in the first chapter, is fully treated in the well-known text-books of L. Silberstein (Macmillan and Co.) and E. Cunningham (Camb. Univ. Press). A useful review of the mathematical theory of Chapter III., giving a fuller account from the standpoint of the pure mathematician, will be found in "Cambridge Mathematical Tracts," No. 9, by J. E. Wright. Finally, for those who wish to learn more of the outstanding discrepancies between astronomical observation and gravitational theory, the following references may be given:

W. DE SITTER. The Secular Variations of the Elements of the Four Inner Planets, "Observatory," XXXVI., p. 296.

E. W. BROWN. The Problems of the Moon's Motion, " Observatory," XXXVII., p. 206.

H. GLAUBERT. The Rotation of the Earth, "Monthly Notices of the Royal Astr. Soc.," LXXV., p. 489.

PREFACE TO SECOND EDITION

The advances made in the eighteen months since this Report was written do not seem to call for any modification in the general treatment. Perhaps the most notable event is the verification of Einstein's prediction as to the deflection of a ray of light by the sun's gravitational field. This was tested at the total eclipse of May 29, 1919, at two stations independently, by expeditions sent out by the Royal and Royal Astronomical Societies jointly, under the superintendence of the Astronomer Royal. The deflection, reduced to the sun's limb should be $1''.75$ on the relativity theory, and $0''.87$ (or possibly zero) according to previous theories. At Principe, where the observations were very much interfered with by cloud, the value $1''.61$ was obtained, with a probable error of $0''.3$; the accuracy appears to be sufficient to indicate fairly decisively Einstein's value. At Sobral, where a clear sky prevailed, the observed value was $1''.98$; the accordance of results derived from right ascensions and declinations, respectively, and the agreement of the displacements of individual stars with the theoretical law demonstrate in a particularly satisfactory manner the trustworthiness of the observations at this station. The full results will be published in a Paper by Sir F. W. Dyson, A. S. Eddington, and C. Davidson in the Philosophical Transactions of the Royal Society.

The test of the displacement of the Fraunhofer lines to the red stands where it did, and we still think that judgment must be reserved. In view of the possibility of a failure in this test, it is of interest to consider exactly what part of the theory can now be considered to rest on a definitely experimental basis. I think it may now be stated that Einstein's law of gravitation is definitely established by observation in the following form:

Every particle and light-pulse moves so that the integral of ds between two points on its track is stationary, where (equation (28.8))

$$ds^2 = -\left(1 - \frac{2m}{r}\right)^{-1} dr^2 - r^2\, d\theta^2 - r^2\, \sin^2\theta\, d\varphi^2 + \left(1 - \frac{2m}{r}\right) dt^2$$

in appropriate polar co-ordinates, the co-efficient of dr^2 being verified to the order m/r, and the co-efficient of dt^2 to the order m^2/r^3. This is checked for high speeds by the deflection of light, and for comparatively low speeds by the motion of perihelion of Mercury, so that unless the true law is of a kind much more complicated than we have allowed for, our expression cannot well be in error.

Accepting Einstein's law in this form, the properties of invariance for transformations of co-ordinates follow, and we reach the conclusion that the intermediary quantity ds (to which as yet we have assigned no physical interpretation) is an invariant, that is to say it has some absolute significance in external nature.

Einstein's theory (as distinct from his *law* of gravitation) gives a physical interpretation to ds, as a quantity that can be measured with material scales and clocks. It is this interpretation which the observation of the Fraunhofer lines should test. The quantity ds is an ideal measure of space and time; and it is possible that we have not yet reached finality as to the right way of realising the ideal practically. It is a fair prediction that an atomic vibration will register ds like an ideal clock; and it is difficult to see how this can be avoided unless the equations of vibration of an atom involve the Riemann-Christoffel tensor. But, if the test fails, the logical conclusion would seem to be that we know less about the conditions of atomic vibration than we thought we did.

A very notable extension of the theory to include electro- magnetic forces and gravitational forces in one geometrical scheme has been given by Prof. H. Weyl in two Papers:

Berlin, Sitzungsberichte, 1918, May 30.
Annalen der Physik, Bd. 59, p. 101.

In Einstein's theory it is assumed that the interval ds has an absolute value, so that two intervals at different points of the world can be immediately compared. In practice the comparison must take place by steps along an intermediate path; for example, by moving a material measuring rod from one point to the other continuously along some path. It is possible that the result of the comparison may not be independent of the path followed, and Weyl considers the electromagnetic field to be the manifestation of this inconsistency. This leads to a very beautiful generalized geometry of the world, in which the electromagnetic field appears as the sign of non-integrability of gauge, and the gravitational field as the sign of non-integrability of direction. The theory has important consequences though it has not suggested any experimental test. It may be added that it appears to favour Einstein's view of the curvature of space, which has been treated, perhaps too unsympathetically, in Chapter VIII.

The writer holds the view that the fundamental equations of gravitation (35.8), which on this theory are the sole basis of mechanics, should be regarded as a definition of matter rather than as a law of nature. We need not suppose that the gravitational field has *in vacuo* some innate tendency to arrange itself according to the law $G_{\mu\nu} = 0$; we should rather say that in regions of the world where this state happens to exist we perceive emptiness; and where the equations fail, the failure of the equations is itself the cause of our perception of matter. Matter does not cause the curvature (G) of space- time; it *is* the curvature. Just as light does not cause electro- magnetic oscillations; it is the oscillations. This point of view is developed in a Paper which will appear shortly in "Mind."

Finally, a word may be added for those who find a difficulty in the combination, of space and time into a static four- dimensional world, in which events do not "happen" they are just there, and we come across them successively in our exploration. "Surely there is a difference between the irrevocable past and the open future, different in quality from the arbitrary distinction of right and left." We agree entirely; but this difference, whatever it is, does not enter into the determinate equations of physics. For physics, the future is $+t$ and the past $-t$, just as right is $+x$ and left $-x$. If we change the place of one particle in our problem we alter the past as well as the future, in contrast to what appears to be the ordinary experience of life, that our interference will alter the future but not the past. The static four-dimensional representation may thus be not completely adequate, but it suffices for all that comes within the purview of physics.

December, 1919.

Contents

1 The Restricted Principle of Relativity

1. In 1887 the famous Michelson-Morley experiment was performed with the object of detecting the earth's motion through the aether. The principle of the experiment may be illustrated by considering a swimmer in a river. It is easily realized that it takes longer to swim to a point 50 yards up-stream and back than to a point 50 yards across-stream and back. If the earth is moving through the aether there is a river of aether flowing through the laboratory, and a wave of light may be compared to a swimmer travelling with constant velocity relative to the current. If, then, we divide a beam of light into two parts, and send one half swimming up the stream for a certain distance and then (by a mirror) back to the starting point, and send the other half an equal distance across-stream and back, the across-stream beam should arrive back first.

Let the aether be flowing relative to the apparatus with velocity u in the direction Ox (Fig. 1); and let OA, OB be the two arms of the apparatus of equal length a, OA being placed up-stream. Let v be the velocity of light. The time for the double journey along OA and back is

$$t_1 = \frac{a}{v-u} + \frac{a}{v+u} = \frac{2av}{v^2-u^2} = \frac{2a}{v}\beta^2 \tag{1.1}$$

where $\beta = (1 - u^2/v^2)^{-1/2}$, a factor greater than unity.

For the transverse journey the light must have a component velocity u up-stream (relative to the aether) in order to avoid being carried below OB; and, since its total velocity is v, its component across-stream must be $\sqrt{v^2-u^2}$. The time for the double journey OB is accordingly

$$t_2 = \frac{2a}{\sqrt{v^2-u^2}} = \frac{2a}{v}\beta, \tag{1.2}$$

so that $t_1 > t_2$

But when the experiment was tried, it was found that both parts of the beam took the same time, as tested by the interference bands produced. It would seem that OA and OB could not really have been of the same length; and if OB was of length a_1, OA must have been of length a_1/β. The apparatus was now rotated through 90°, so that OB became the up-stream arm. The time for the two journeys was again the same, so that OB must now be the shorter arm. The plain meaning of the experiment is that both arms have a length a_1 when placed along Oy, and automatically contract to a length a_1/β when placed along Ox. This explanation was first given by FitzGerald.

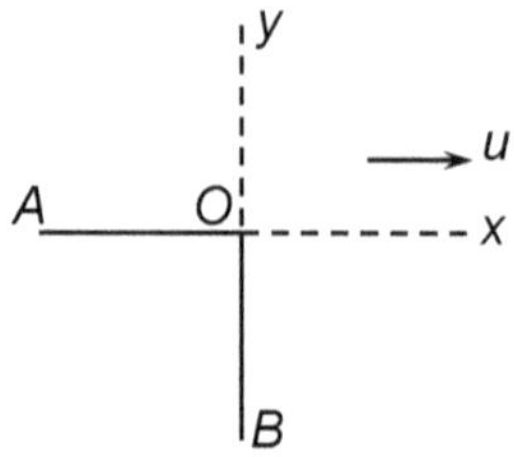

Fig. 1

It is not known how much the earth's motion through the aether amounts to; but at some time during the year it must be at least 30 km. per sec., since the earth's velocity changes by 60 km. per sec. between opposite seasons. The experiment would have detected a velocity much smaller than this (about 3 km. per sec.), if it were not for the compensating contraction of the arms of the apparatus. By experimenting at different times of the year with different orientations the existence of the contraction has been fully demonstrated. It has been shown that it is independent of the material used for the arms, and the contraction is in all cases measured by the ratio $\beta = (1 - u^2/v^2)^{-1/2}$.

It is now known that this contraction fits in well with the electrical theory of matter, and may be attributed to changes in the electromagnetic forces between the particles which determine the equilibrium form of a so-called rigid body. This universal property of matter is therefore not so mysterious as it at first seemed; and we shall not here discuss the unsuccessful attempts at alternative explanations of the Michelson-Morley experiment, i.e., by assuming a convection of the aether by the earth.

2. The Michelson-Morley experiment has thus unexpectedly failed to measure our motion through the aether, and many other ingenious experiments have failed in like manner. So far as we can test, the earth's motion makes absolutely no difference in the observed phenomena; and we shall not be led into any contradiction with observation if we assign to the earth any velocity through the aether that we please. It is interesting to trace in a general way how this can happen. Let us assign to the earth a velocity of 161,000 miles a second, say, in a vertical direction. With this speed $\beta = 2$, and the contraction is one-half. A rod 6 feet long when horizontal contracts to 3 feet when placed vertically. Yet we never notice the change. If the standard yard-measure is brought to measure it, the rod will still be found to measure two yards; but then the yard-measure experiences the same contraction when placed alongside, and represents only half-a-yard in that position. It might be thought that we ought to see the change of length when the rod is rotated. But what we perceive is an image of the rod on the retina of the eye; we think that the image occupies the same space of retina in both positions; but our retina has contracted in the vertical direction without our knowing it, and our estimates of length in that direction are double

what they should be. Similarly with other tests. We might introduce electrical and optical tests, in which the cause of the compensation is more difficult to trace; but, in fact, they all fail. The universal nature of the change makes it impossible to perceive any change at all.

3. This discussion leads us to consider more carefully what is meant by the *length* of an object, and the *space* which we consider it to occupy. To the physicist, space means simply a scaffolding of reference, in which the mind instinctively locates the phenomena of nature. Our present point of view assumes that there is a "real " or "absolute" scaffolding, in which a material body moving with the earth changes its length according as it is oriented in one direction or another. On the other hand, the human race (and its predecessors) have conceived and used a different scaffolding – the space of appearance – in which a material body moving with the earth does not change length as its orientation alters. It often happens that a primitive conception is ambiguous, and has to be re-defined when adopted for scientific purposes; but there is little justification for doing this in the case of space. Firstly, the space of appearance is perfectly suitable for scientific purposes, since we have just seen that it is impossible to detect experimentally that it is not the absolute space. Secondly, so long as we cannot detect our motion through the aether, we do not know how to convert our observations so as to express them in terms of absolute space. Thirdly, for all we know, our velocity through the aether may be so great that the absolute space and the space of appearance do not even approximately correspond; thus we might be revolutionizing rather than re-defining the common conception of space.

It will therefore be considered legitimate to use the words "space" and "length" with their current significance. A rigid body on the earth is generally considered not to change length when its direction is altered, and by this property we block out a scaffolding of reference for our measures and locate objects in *our* space – the space of appearance. But we have learnt one important thing. Our space is not absolute; it is determined by our motion. If we transfer ourselves to the star Arcturus, which is moving relatively to us with a speed of more than 300 km. per sec., our space will not suit it, since it was designed to eliminate our own contraction effects. The contraction ratio β must be different for Arcturus; and the space surveyed with a material yard-measure carried on Arcturus will differ slightly from the space surveyed with the same yard-measure on the earth. It may also be noted that there is a slight difference in our own space in summer and winter (owing to the change of the earth's motion), and this may have to be taken into account in some applications.

Accordingly by "space" we shall mean the space of appearance for the observer considered. It becomes definite when we specify the motion of the observer. In particular, if the observer is at rest in the aether, the corresponding space is what we have hitherto called the "absolute space."

The possibility of different observers using different spaces may be illustrated by considering the question, What is a circle? Suppose a circle is drawn on paper in the usual way with a pair of compasses. An observer S, who believes the paper to be moving through the aether with a great velocity, must, in accordance with the Michelson-Morley experiment, suppose that the distance between the points of the compasses changed as the curve was described; he will therefore deem the curve to be an ellipse. Another observer S' who believes the paper to be at rest in the aether, will deem it to be a circle. There is no experimental means of finding out which is right in his hypothesis. We have, therefore, to admit that the same curve may be arbitrarily regarded as an ellipse or as a circle. That illustrates our meaning when we say that S and S' use different spaces, the curve being an ellipse in one space and a circle in the other

4. The failure of all experimental tests to decide whether the space of S or of S' is the more fundamental is summed up in the restricted Principle of Relativity. This asserts that *it is impossible by any conceivable experiment to detect uniform motion through the aether.* This generalization is based on a great amount of experimental evidence, which is fully discussed in text-books on the older theory of relativity. Here it is perhaps sufficient to state that experimental confirmation appears to be sufficient,[1] except in regard to the question whether gravitation falls within the scope of the principle. We shall assume that the principle is true universally.

Let x', y', z', be the co-ordinates of a point in the space of an observer S'; and let x, y, z be the co-ordinates of the same point in the space of an observer S at rest in the aether. Let S' move relatively to S with velocity u in the direction Ox. S', using his own space, has no knowledge of his motion through the aether, and he makes all his theoretical calculations as though he were at rest; from what has been already said, he will not discover any contradiction with observation.

According to ordinary kinematics the relation between the co-ordinates and the times (t', t) in the two systems would be

$$x' = x - ut, \quad y' = y, \quad z' = z, \quad t' = t \tag{4.1}$$

But the first of these must be modified, because in the $x-$direction S''s standard of length is contracted in the ratio $\frac{1}{\beta}$ The equation becomes

$$x' = \beta(x - ut) \tag{4.15}$$

In order to satisfy the principle of relativity, it appears that the time t' used by S' must differ from the time t used by S.

We shall suppose that both observers use the same value for the velocity of light; this is merely a matter of co-ordinating their units, the significance of which will be considered in the next paragraph. Let S' observe the time t' taken for the double journey $OB = 2a_1$ in Fig. 1. It must agree with his calculated time, which is, of course, $\frac{2a_1}{v}$. Thus

$$t' = \frac{2a_1}{v}.$$

But in (1.2), when we were using S's co-ordinates, we found the time to be

$$t = \frac{2a_1\beta}{v}.$$

[1] *I.e.*, sufficient to assert the *universality* not necessarily the perfect *accuracy*, of the principle.

Hence

$$t = \beta t'.$$

This also fits the double journey OA. S', unaware of his motion, does not allow for any contraction, and calculates the time for the double journey as

$$t' = \frac{2a_1}{v}.$$

But S recognizes the contraction, and considers the distance travelled to be $\frac{2a_1}{\beta}$. Hence calculating as in (1.1), he makes the time to be

$$t = \frac{2a_1}{\beta v}\beta^2,$$

so that again $t = \beta t'$.

Accordingly S' must use a unit of time longer than that of S in the ratio β; otherwise he would find a discrepancy between observation and calculation.

There is another difference in time-measurement involved. According to S, the light completes the half-journey OA in a time $\frac{a_1/\beta}{v-u}$ in S's units, or $\frac{a_1/\beta^2}{v-u}$ in S' units of time. But

$$\frac{a_1}{\beta^2(v-u)} = \frac{a_1(v+u)}{v^2} = \frac{a_1}{v} + \frac{a_1 u}{v^2}.$$

But the difference in the time of leaving O and reaching A must be deemed by S' to be $\frac{a_1}{v}$; he must therefore set his clock at A $\frac{a_1 u}{v^2}$ slow compared with the clock at O. He has no idea that it is slow; he has attempted to adjust the two clocks together. But his determination of simultaneity of events at O and A differs from that of S, because he allows a different correction for the time of transit of the light.

Including both these differences, we see that the relation between the times adopted by S and S' is

$$t = \beta\left(t' + \frac{ux'}{v^2}\right).$$

Substituting this value of t in (4.15) we obtain after an easy reduction

$$x = \beta(x' + ut').$$

Collecting together our results, we have the formulae of transformation

$$x = \beta(x' + ut'), \quad y = y', \quad z = z', \quad t = \beta\left(t' + \frac{ux'}{v^2}\right). \tag{4.2}$$

By the principle of relativity nothing is altered if S is in motion relative to the aether; so the relations (4.2) must hold between the spaces and times of *any* two observers having relative velocity u.

By solving (4.2) for x', y', z', t', we obtain the reciprocal relations

$$x' = \beta(x - ut), \quad y' = y, \quad z' = z \quad t' = \beta\left(t - \frac{ux}{v^2}\right). \tag{4.3}$$

These might have been written down immediately, because interchanging S and S' is equivalent to reversing the sign of u; but it will be seen later that the verification by direct solution of (4.2) is important.

5. We have supposed that S and S' adopt the same measure for the velocity of light; this was in order to secure that the units of velocity used by S and S' correspond. It is no use for S to describe his experiences to S' in terms of units which are outside the knowledge of the latter; but if S states that a velocity occurring in his experiment is a certain fraction of the velocity of light S' will be able to compare that with his own experimental results. By the principle of relativity any other velocities occurring in their experiments under similar conditions will correspond; and, for example, we see from (4.2) and (4.3) that they will agree in calling their relative velocity $+u$ and $-u$ respectively.

Whilst this settles the consistency of the units of velocity used in (4.2) we have not yet secured that the units of length correspond. A description of Brobdingnag by a Brobdingnagian would not have mentioned the most striking feature of that country; it needed an intruding Gulliver to detect the enormous scale of everything contained. And so we may ask whether a natural standard of length, say a hydrogen atom, at rest in S's system will be of the same size in terms of x, y, z, as a hydrogen atom at rest in S' system in terms of x', y', z', Clearly it will be misleading if we do not correlate the co-ordinates so as to satisfy this.

To allow for a possible non-correspondence of the units of length in (4.2) we can write the transformation more generally

$$kx = \beta(x' + ut'), \quad ky = y' \quad , kz = z', \quad , kt = \beta(t' + \frac{ux'}{v^2}), \tag{5.2}$$

where k depends on the magnitude, but clearly not on the direction, of u.

But now applying (5.2) the reverse way, i.e., regarding x, y, z, t as a system moving with velocity $-u$ relative to x', y', z', t' we shall have

$$kx' = \beta(x - ut), \quad ky' = y, kz' = z, \quad , kt' = \beta(t - \frac{ux}{v^2}), \tag{5-3}$$

which is clearly inconsistent with (5.2) unless $k = 1$. Hence (4.2) gives the only possible correspondence of the units of length.

We thus use the remarkable property of reciprocity possessed by (4.2) and (4.3), but not by (5.2) and (5.3), to fix the necessary correspondence of the units. The dimensions of a motionless hydrogen atom will now be the same in both systems; for, if not, we could find a system in which the dimensions were either a maximum or a minimum; and that system would give us an absolute standard from which we could measure absolute motion.

It is thus clear that S' will actually measure his space and time by the variables $x'y', z', t'$ given by (4.3), if he sets about choosing his units in the same way that S did.

6. We have established the connection between the co-ordinates used by S and S' by reference to simple criteria. It is interesting to work out in detail the correspondence of the two systems for other and more complex phenomena,

showing that the transformation always works consistently. But the standpoint of the principle of relativity rather discourages this procedure. Its view is that the indifference of all natural phenomena to an absolute translation is something immediately understandable, whilst the contractions and other complications entering into our description arise from our own perversity in not looking at Nature in a broad enough way. When a rod is started from rest into uniform motion, nothing whatever happens to the rod. We say that it contracts; but length is not a property of the rod; it is a *relation* between the rod and the observer. Until the observer is specified the length of the rod is quite indeterminate. We ought always to remember that our experiments reveal only relations, and not properties inherent in individual objects; and then the correspondence of two systems, differing only in uniform motion, becomes axiomatic, so that laborious mathematical verifications are redundant. Human minds being what they are, that is a counsel of perfection, and we shall not follow it too strictly.

The only verification that is needed is to show that our fundamental laws of mechanics and electrodynamics are consistent with the principle of relativity. This will be done in connection with a much more general principle of relativity for mechanics in §37, and for electrodynamics in §45.

7. (a) As an illustration of the modification of ordinary views required by this theory, we may notice the law of composition of velocities. Consider a particle moving relative to S with velocity w along Ox, so that

$$\frac{dx}{dt} = w. \tag{7.1}$$

The velocity relative to S' will be

$$\begin{aligned} w' = \frac{dx'}{dt'} &= \frac{\beta(dx - udt)}{\beta(dt - udx/v^2)} \quad \text{by (4.3)}, \\ &= \frac{w-u}{1-uw/v^2} \quad \text{by (7.1)}. \end{aligned} \tag{7.2}$$

The velocity relative to S' is thus not $w - u$, as we should have assumed in ordinary mechanics.

It has been pointed out by Robb that the addition-law for motion in one dimension can be restored if we measure motion by the *rapidity*, $\tanh^{-1}/(w/v)$ instead of by the velocity w. Equation (7.2) gives

$$\tanh^{-1}\left(\frac{w'}{v}\right) = \tanh^{-1}\left(\frac{w}{v}\right) - \tanh^{-1}\left(\frac{u}{v}\right). \tag{7.3}$$

Since $\tanh^{-1} 1 = \infty$, the velocity of light corresponds to infinite rapidity, and we may compound any number of relative velocities less than that of light without obtaining a resultant greater than the velocity of light,

(b) To find the relation of the densities σ = number of particles per unit volume) in the two systems, we can easily verify that the Jacobian $\frac{\partial(x', y', z' t')}{\partial(x, y, z, t)} = 1$, so that

$$dx' dy' dz' dt' = dx\, dy\, dz\, dt. \tag{7.4}$$

But the number of particles in a particular element of volume cannot depend on the co-ordinates used to describe the element, hence

$$\sigma' dx' dy' dz' = \sigma \, dx \, dy \, dz. \tag{7-5}$$

Hence

$$\frac{\sigma'}{\sigma} = \frac{dt'}{dt} = \beta\left(1 - \frac{uw}{v^2}\right) \tag{7.6}$$

since $dx/dt = w$.

In particular, if $w = 0$, so that σ is the density referred to axes moving with the matter,

$$\sigma' = \beta\sigma. \tag{7-65}$$

Since the mass of a particle may depend on its motion, we cannot assume that the ratio ρ'/ρ of the mass-density is the same as that of the distribution-density σ'/σ

When the transformation (4.2) was first introduced in electro-dynamics by Larmor and Lorentz, t' was regarded as a fictitious time introduced for mathematical purposes, and it was scarcely realized that it was the actual measured time of the moving observer. Einstein in 1905 first showed that velocity and density would be estimated by the moving observer in the way given above, and thus removed the last discrepancy between the electrodynamical equations for the two systems.

(c) In order to find the change (if any) of mass with velocity, consider a body of mass m_1, m_1' (in the two systems of reference) moving with velocity w_1, w_1'. Let

$$\beta_1 = \left(1 - \frac{w_1^2}{v^2}\right)^{-\frac{1}{2}}, \quad \beta_1' = \left(1 - \frac{{w_1'}^2}{v^2}\right)^{-\frac{1}{2}}.$$

Working out β_1' by using (7.2), we easily find

$$\beta_1' w_1' = \beta\beta_1(w_1 - u). \tag{7.71}$$

Let a number of bodies be moving in a straight line subject to the conservation of mass and momentum, i.e.,

$$\Sigma m_1 \quad \text{and} \quad \Sigma m_1 w_1 \quad \text{are conserved.}$$

Then, since u and β are constants,

$$\beta \, \Sigma m_1(w_1 - u) \quad \text{will be conserved.}$$

Therefore by (7.71)

$$\Sigma \frac{m_1 \beta_1'}{\beta_1} w_1' \quad \text{is conserved.} \tag{7.72}$$

But since momentum must be conserved for the observer S'

$$\Sigma m_1' w_1' \quad \text{is conserved.} \tag{7-73}$$

The results (7.72) and (7.73) will agree if

$$\frac{m_1}{\beta_1} = \frac{m_1'}{\beta_1'} = m_0, \quad \text{say},$$

and it is easy to show that there is no other solution. Hence

$$m_1 = m_0\beta_1 = m_0\left(1 - \frac{w_1^2}{v^2}\right)^{-\frac{1}{2}}, \tag{7.8}$$

where m_0 is constant and equal to the mass at rest. This is the law of dependence of mass on velocity.

Neglecting w_1^4/v^4, we have

$$m_1 = m_0 + \frac{\frac{1}{2}m_0 w_1^2}{v^2} \cdots \tag{7.85}$$

so that we may regard the mass as made up of a constant mass m_0 belonging to the particle, together with a mass proportional to, and presumably belonging to, the kinetic energy. If we choose units so that the velocity of light is unity, the mass of the energy is the same as the energy, and the distinction between energy and mass is obliterated. Accordingly m_0 is also regarded as a form of energy. (It is usually identified mainly with the electrostatic energy of the electrons forming the body.)

Since the conservation of mass now implies the conservation of energy we have to restrict the reactions between the bodies in the foregoing discussion to perfectly elastic impacts. Other interactions would require a more general treatment; in fact, if the energy is not conserved, the momentum is not perfectly conserved, because the disappearing energy has mass and therefore carries off momentum.

In this discussion we are justified in pressing the laws of conservation of mass and momentum to the utmost limit as holding with absolute accuracy, since the definition and measurement of mass (inertia) rests on these laws,[2], and unless we have an accurate definition it is meaningless to investigate change of mass. In astronomy, however, the masses of heavenly bodies are measured by their gravitational effects; naturally we cannot legitimately apply (7.8) to *gravitational mass* without a full discussion of the law of gravitation.

It should be noticed that this change of mass with velocity is in no way dependent on the electrical theory of matter.

(d) To find the transformation of mass-density ρ, we have

$$\frac{\rho'}{\rho} = \frac{\sigma' m'}{\sigma m} = \frac{\sigma'\beta_1'}{\sigma\beta_1} = \frac{\beta_1'\beta}{\beta_1}\left(1 - \frac{uw_1}{v^2}\right) \quad \text{by (7.6)},$$

which becomes by (7.71)

$$\frac{\rho'}{\rho} = \left(\frac{\beta_1'}{\beta_1}\right)^2. \tag{7-91}$$

[2] The mass here discussed is sometimes called the "transverse mass." The so-called longitudinal mass is of no theoretical importance, it is not conserved, it does not enter into the expression for the momentum or energy, and it has no connection with gravitation.

In particular, if ρ_0 is the density in natural measure, i.e., referred to axes moving with the matter, ρ the density referred to axes with respect to which the matter has a velocity u,

$$\rho = \beta^2 \rho_0. \tag{7-92}$$

8. Of late years the domain of the electromagnetic theory has been extended, so that most natural phenomena are now attributed to electrical actions. The relativity theory does not presuppose an electromagnetic theory either of matter or of light; but, if we accept the latter theories, it becomes possible to state exactly the points on which experimental evidence is required in order to establish our hypothesis. The experimental laws of electromagnetism are summed up in Maxwell's equations; and in so far as these cover the phenomena, the complete equivalence of the sequence of events in a fixed system described in terms of x, y, z, t, and a moving system described in terms of x', y', z', t', has been established analytically. So far as is known, only three kinds of force are outside the scope of Maxwell's equations.

(1) The forces which constrain the size and shape of an electron are not recognised electromagnetic forces. Fortunately the properties of an electron at rest and in extremely rapid motion can be studied experimentally, and it is believed that they change in the way required by relativity.

(2) The phenomena of Quanta appear to obey laws outside the scope of Maxwell's equations. Theoretically these laws fit in admirably with relativity, since Planck's fundamental unit of action is found to be unaltered by the choice of axes. But on the experimental side, evidence of the relativity of phenomena involving quantum relations has not yet been produced. This is particularly unfortunate, because the vibration of an atom depends on quantum relations; and it is practically essential to the relativity theory that an atom (acting as a natural clock) should keep the time appropriate to the axes chosen.

(3) Gravitation is outside the electromagnetic scheme. The Michelson-Morley experiment is necessarily confined to solids of laboratory dimensions, in which internal gravitation has no appreciable influence. There is, therefore, no experimental proof that a body such as the earth, whose figure is determined mainly by gravitation, will undergo the theoretical contraction owing to motion. The most direct evidence that gravitation conforms to relativity comes from a discussion by Lodge[3] of the effect of the sun's motion through the aether on the perihelia and eccentricities of the inner planets. If gravitation is outside the relativity theory (the Newtonian law holding unmodified) a solar motion of 10 km. per sec. would produce perturbations in the eccentricities and perihelia of the earth and Venus, which could probably be detected by observation. The absence of these perturbations seems to show that gravitation must conform to relativity, unless, indeed, the sun happens to be nearly at rest in the aether. If we confine attention to our local stellar system the average stellar velocities are not so much greater than 10 km. per sec. as to render the latter alternative too improbable; but the very high velocities found for the spiral nebulae (which are thought to be distant stellar systems) makes it improbable that our local system should be

[3]Phil. Mag., February, 1918.

so nearly at rest in the aether.

2 The Relations of Space, Time, and Force

9. An interesting aspect of the transformation of the variables x, y, z, t to x', y', z', t' has been brought out by Minkowski. We consider them as co-ordinates in a four-dimensional continuum of space and time. Choose the units of space and time so that the velocity of light is unity, and set

$$t = i\tau, \quad \text{where} \quad i = \sqrt{-1}.$$

The equations of transformation (4.2) become

$$\begin{gathered} x = \beta(x' + iu\tau'), \quad y = y', \quad z = z' \quad \tau = \beta(\tau' - iux') \\ \beta = (1 - u^2)^{-\frac{1}{2}} \end{gathered} \tag{9.1}$$

Let $u = i \tan\theta$, so that θ is an imaginary angle. Then $\beta = \cos\theta$, and (9.1) becomes

$$x = x'\cos\theta - \tau'\sin\theta, \quad y = y', \quad z = z', \quad \tau = \tau'\cos\theta + x'\sin\theta. \tag{9.2}$$

Thus the transformation is simply a rotation of the axes of co-ordinates through an imaginary angle θ in the plane of $x\tau$.

We know that the orientation chosen for the space-axes, x, y, z, makes no difference in Newtonian mechanics. The principle of relativity extends this so as to include the axis τ. The continuum formed of space and imaginary time is perfectly isotropic; the resolution into space and time separately, which depends on the motion of the observer, corresponds to the arbitrary orientation in it of a set of rectangular axes.

10. From this point of view the strange conspiracy of the forces of Nature to prevent the detection of our absolute motion disappears. There is no conspiracy of concealment, because there is nothing to conceal. The continuum being isotropic, there is no orientation more fundamental than any other; we cannot pick out any direction as the absolute time any more than we can pick out a direction in space as the absolute vertical. Up-and-down, right-and-left, backwards-and-forwards, sooner-and-later[1] equally express relations to

[1]This applies to imaginary time. With real time, events which (as usually happens) are

some particular observer, and have no absolute significance. In Minkowski's famous words, "Henceforth Space and Time in themselves vanish to shadows, and only a kind of union of the two preserves an independent existence."

The scientific basis of the idea that some fundamental division into space and time exists was the conception of the aether as a material fluid, filling uniformly and isotropically a particular space. It now seems clear that the aether cannot have those material properties which would enable it to serve as a frame of reference. Its functions seem to be limited to those summed up in the old description "the nominative of the verb 'to undulate'. "

Unfortunately the simplicity of this conception of the four-dimensional continuum is only formal; and natural phenomena make a discrimination between τ and the other variables by relating themselves to an imaginary τ, which we call the time. In natural variables, x, y, z, t this view of the transformation as a rotation of axes becomes concealed.[2]

11. In the four-dimensional continuum the interval δs between two point-events is given by

$$-\delta s^2 = \delta x^2 + \delta y^2 + \delta z^2 + \delta \tau^2, \qquad (11.1)$$

which is unaffected by any rotation of the axes, and is therefore invariant for all observers. The minus sign given to δs^2 is an arbitrary convention, and the formula is simply the generalization of the ordinary equation

$$\delta s^2 = \delta x^2 + \delta y^2 + \delta z^2.$$

The fact that δs is measured consistently by all observers who would obtain discordant results for δx, δy, δz, $\delta \tau$ separately, is so important in our subsequent work that we shall consider the nature of the clock-scale needed for its measurement.

We have a scale AB divided into kilometers, say, and at each division is placed a clock also registering kilometers. (The velocity of light being unity, a kilometer is also a unit of time $\frac{1}{300000}$ sec.). When, the clocks are correctly set and viewed from A, the sum of the readings of any clock and the division beside it is the same for all, since the scale-reading gives the correction for the time taken by light in travelling to A. This is shown in Fig. 2, where the clock-readings are given as though they were being viewed from A.

Now lay the scale in line with the two events; note the clock and scale-reading, t_1, σ_1 of the first event, and the corresponding readings $_2$, σ_2 of the second event; then from (11.1)

$$\delta s^2 = (t_2 - t_1)^2 - (\sigma_2 - \sigma_1)^2. \qquad (11.2)$$

If the scale had been set in motion in the direction AB, $\sigma_2 - \sigma_1$ would have been diminished, owing to the divisions having advanced to meet the second

separated by a greater interval in time than in space preserve the same order for all observers. But an event on the sun which we should describe as occurring 2 minutes later than an event on the earth might be described by another observer as 2 minutes earlier. (Both observers have corrected their observations for the light-time.)

[2]For a logical study of the properties of the continuum of space and real time reference may be made to A. A. Robb, "A Theory of Time and Space" (Camb. Univ. Press).

event. But the clocks would have been adjusted differently, because A is now advancing to meet the light coming from any clock, and the clock would appear too fast (by the above rule) if it were not set back. There are other second-order corrections arising from the contraction of the scale and change of rate of the clocks owing to motion; but the net result is a perfect compensation, and δs^2 determined from (11.2) must be invariant, as already proved.

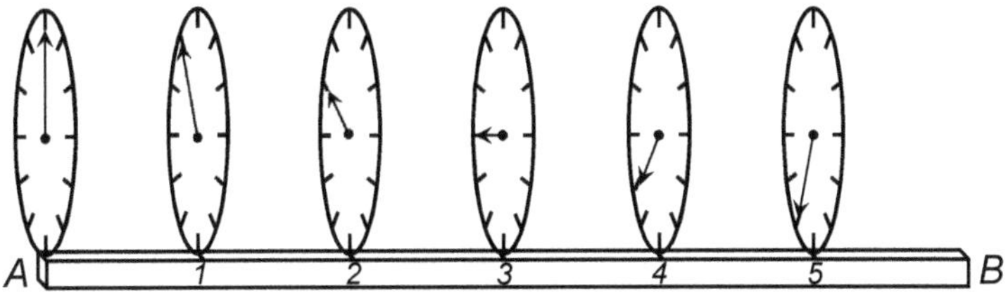

Fig. 2

It is clear that the whole (restricted) principle of relativity is summed up in this invariance of δs, and it is possible to deduce the equation of transformation (4.2) and our other previous results by taking this as postulate.

When δs refers to the interval between two events in the history of a particular particle it has a special interpretation which deserves notice. If we choose axes moving with the particle, δx, δy, $\delta z = 0$, so that $\delta s = \delta t$. Accordingly the variable s is called the "proper-time," i.e., the time measured by a clock attached to the particle.

12. Up to the present we have discussed a particular type of transformation of co-ordinates, viz., that corresponding to a uniform motion of translation. We now enter on the theory of more general changes of co-ordinates.

The co-ordinates x, y, z, t of a particle trace a curve in four dimensions which is called the *world-line* of the particle. If we draw the world-lines of all the particles, light-waves and other entities, we obtain a complete history of the configurations of the Universe for all time. But such a history contains a great deal that is necessarily outside experience. All exact observations are records of coincidences of two entities in space and time, that is to say, records of intersections of world-lines.

It is easy to see that this is the case in laboratory experiments or astronomical observations. Electrical measurements, determinations of temperature, weight, pressure, etc., rest finally on the coincidence of some indicator with a division on a scale. Many of our rough observations depend on coincidences of light waves with elements of the retina, or the simultaneous impact of sound-waves on the ear. It is true that some of our external knowledge is not obviously of this character. We estimate the weight of a letter, balancing it in the hand; this is based on a muscular sensation having no immediate relation to time and space, but we fit this crude knowledge into the exact scheme of physics by comparing it with more accurate measures based on coincidences.

The observation that the world-lines of two particles intersect is a genuine addition to knowledge, since in general lines in space of three or four dimensions miss one another. We have to build up our conception of the location of objects in space and time from a large number of records of coincidences. It is clear that we have a great deal of liberty in drawing the world-lines, whilst satisfying all the intersections. Let us draw the world-lines in some admissible way, and imagine them embedded in a jelly. If the jelly is distorted in any way, the world-lines in their new courses will still agree with observation, because no intersection is created or destroyed.

Mathematically this can be expressed by saying that we may make any mathematical transformation of the co-ordinates. If we choose new co-ordinates x', y', z', t', which are any four independent functions of x, y, z, t, a coincidence in x, y, z, t will also be a coincidence in x', y', z', t', and *vice versa*. By locating objects in the space-time given by x', y', z', t', we do not alter the course of events. The events themselves do not presuppose any particular system of co-ordinates, and the space-time scaffolding is something introduced arbitrarily by ourselves.

It is almost a truism to say that we may adopt any system of co-ordinates we please. We are accustomed to introduce curvilinear co-ordinates or moving axes without apology, whenever they simplify the problem. But there is one point not so generally recognized. Ordinarily when we use curvilinear co-ordinates we never allow ourselves to forget that they *are* curvilinear; it is a mathematical device, not a new space, that we adopt. Perhaps the only case in which we really take the new co-ordinates seriously is in the transformation to rotating axes; we then take account of the rotation by adding a fictitious centrifugal force to the equations, and thenceforth the rotation is quite put out of mind. From the standpoint of relativity, when we adopt new co-ordinates x', y', z', t' we shall adopt a corresponding new space, and think no more of the old space. For instance, a "straight line" in the new space will be given by a linear relation between x', y', z', t'.

The behaviour of natural objects will no doubt appear very odd when referred to a space other than that customarily used. So-called rigid bodies will change dimensions as they move; but we are prepared for that by our study of the Michelson- Morley contraction. Paths of moving particles will for no apparent reason deviate from the "straight line" but, accepting the definition of a force as that which changes a body's state of rest or motion, this must be attributed to a field of force inherent in the new space (cf. the centrifugal force). Light-rays will also be deflected, so that the field of force acts on light as well as on material particles, this is not altogether a novel idea, because a little reflection shows that the centrifugal force deflects light as well as matter although optical problems are not usually treated in that way.

13. The laws of mechanics and electrodynamics are usually enunciated with respect to "unaccelerated rectangular axes" or, as they are often called, "Galilean axes." We cannot regard such axes as recognisable intuitively, and the only definition of them that can be given is that they are the axes with respect to which that particular form of the laws holds. It is part of the method of the present theory to restate the laws of Nature in a form not confined to Galilean co-ordinates, so that all systems of co-ordinates are regarded as on the same footing.

In unaccelerated rectangular co-ordinates the path of a particle is a straight line (apart from the influence of other matter, or the electromagnetic field). When we transform to other co-ordinates the path is no longer straight, i.e., it is no longer given by a linear relation between the co-ordinates; and the bending of the path is attributable to a field of force which comes into existence in the new space. This field of force has the property that the deflection produced is independent of the nature of the body acted on, being a purely geometrical deformation. Now the same property is shared by the force of gravitation; the acceleration produced by a given gravitational field is independent of the nature or mass of the body acted on. This has led to the hypothesis that gravitation may be of essentially the same nature as the geometrical forces introduced by the choice of co-ordinates.

This hypothesis, which was put forward by Einstein, is called the Principle of Equivalence. It asserts that *a gravitational field of force is exactly equivalent to a field of force introduced by a transformation of the co-ordinates of reference, so that by no possible experiment can we distinguish between them.*

In Jules Verne's story, "Round the Moon" three men are shot up in a projectile into space. The author describes their strange experiences when gravity vanishes at the neutral point between the earth and moon. Pedantic criticism of so delightful a book is detestable; yet perhaps we may point out that, for the inhabitants of the projectile, weight would vanish the moment they left the cannon's mouth. They and their projectile are falling freely all the time at the same rate, and they can feel no sensation of weight. They automatically adopt a new space, referred to the walls and fixtures of their projectile instead of to the earth. Their axes of reference are accelerated falling towards the earth; and this transformation of axes introduces a field of force which just neutralizes the gravitational field. But, whilst they could detect no gravitational field by ordinary tests, it is not obviously impossible for them to detect some effect by optical or electrical experiments. According to the principle of equivalence, however, no effect of any kind could be detected inside the projectile; the gravitational field cannot be differentiated from a transformation of co-ordinates, and therefore the same transformation which neutralizes mechanical effects neutralizes all other effects.

It will be seen that this principle of equivalence is a natural generalization of the principle of relativity. An occupant of the projectile can only observe the *relations* of the bodies inside to himself and to each other. The supposed absolute acceleration of the projectile is just as irrelevant to the phenomena as a uniform translation is. The mathematical space-scaffolding of Galilean axes, from which we measure it, has no real significance. If the projectile were not allowed to fall, gravity would be detected – or rather the force of constraint which prevents the fall would be detected. I think it is literally true to say that we never feel the force of the earth's attraction on our bodies; what we do feel is the earth shoving against our feet.

14. A limitation of the Principle of Equivalence must be noticed. It is clear that we cannot transform away a natural gravitational field altogether. If we could, we should unconsciously make the transformation and adopt the new co-ordinates just as the inhabitants of the projectile did. They were concerned with

a practically infinitesimal region, and for an infinitesimal region the gravitational force and the force due to a transformation correspond; but we cannot find any transformation which will remove the gravitational field throughout a finite region. It is like trying to paste a fiat sheet of paper on a sphere, the paper can be applied at any point, but as you go away from the point you soon come to a misfit. For this reason it will be desirable to define the exact scope of the principle of equivalence. Up to what point are the properties of a gravitational field and a transformation field identical? And what properties does a gravitational field possess which cannot be imitated by a transformation? The impossibility of transforming away a gravitational field is, of course, an experimental property; so that, in spite of the principle of equivalence, there is at least one means of making an experimental distinction.

Space-time in which there is no gravitational field which cannot be transformed away is called *homaloidal.* In homaloidal space-time then, we can choose axes so that there is no field of force anywhere. Remembering that we have no means of defining axes except from the form of the laws of Nature referred to them, we should naturally take these axes as fundamental and name them "rectangular and unaccelerated." The dynamics of homaloidal space would not recognize the existence of gravitation. Our space is not like that, though we believe that at great distances from all gravitating matter it tends towards this condition as a limit. The necessary limitation of the principle of equivalence turns on the number of consecutive points for which gravitational space-time agrees with homaloidal space-time; in other words, the equivalence will hold only up to a certain order of differential coefficients. Properties involving differential coefficients up to this order will be the same in the gravitational field as in a homaloidal field; whilst properties of the transformed field involving differential coefficients of higher order will not necessarily hold in the gravitational field

The determination of the order of the differential coefficients for which agreement is possible must be deferred to §27. Meanwhile it may be noted that we can always choose axes for which the field at a given *point* vanishes – viz., take rectangular axes moving with the acceleration at that point. In that case we are said to use "natural measure."

15. At a point of space where there is no field of force the observer's clock-scale, if unconstrained, will be either at rest or in uniform motion. We have seen that the measured interval, δs, between two events is independent of uniform motion, and hence a unique value of δs is determined by the measures.

Using rectangular co-ordinates, the relation between an infinitesimal *measured* interval ds and the *inferred* co-ordinates of the event is (11.1).

$$ds^2 = -dx^2 - dy^2 - dz^2 + dt^2. \tag{15.1}$$

Introduce new co-ordinates x_1, x_2, x_3, x_4 which are any functions of x, y, z, t given by

$$x = f_1(x_1, x_2, x_3, x_4), \quad y = f_2(x_1, x_2, x_3, x_4), \text{ etc.}$$

Then

$$dx = \frac{\partial f_1}{\partial x_1}dx_1 + \frac{\partial f_1}{\partial x_2}dx_2 + \frac{\partial f_1}{\partial x_3}dx_3 + \frac{\partial f_1}{\partial x_4}dx_4, \text{ etc.} \tag{15.2}$$

Substituting (15.2) on the right-hand side of (15.1), we obtain a general quadratic function of the infinitesimals, which may be written,

$$\begin{aligned} ds^2 &= g_{11}dx_1^2 + g_{22}dx_2^2 + g_{33}dx_3^2 + g_{44}dx_4^2 \\ &+ 2g_{12}dx_1dx_2 + 2g_{13}dx_1dx_3 + 2g_{14}dx_1dx_4 \\ &+ 2g_{23}dx_2dx_3 + 2g_{24}dx_2dx_4 + 2g_{34}dx_3dx_4 \end{aligned} \tag{15.3}$$

where the g's are functions of the co-ordinates, depending on the transformation.

As an illustration we may take the transformation to rotating axes

$$\begin{aligned} x &= x_1 \cos\omega\, x_4 - x_2 \sin\omega\, x_4 \\ y &= x_1 \sin\omega\, x_4 + x_2 \cos\omega\, x_4 \\ z &= x_3 \\ t &= x_4 \end{aligned} \tag{15.4}$$

Whence

$$\begin{aligned} dx &= \cos\omega\, x_4\, dx_1 - \sin\omega\, x_4\, dx_2 - \omega\,(x_1 \sin\omega\, x_4 + x_2 \cos\omega\, x_4)dx_4 \\ dy &= sin\omega\, x_4\, dx_1 + \cos\omega\, x_4\, dx_2 + \omega\,(x_1 \cos\omega\, x_4 - x_2 \sin\omega\, x_4)dx_4 \\ dz &= dx_3, \quad dt = dx_4. \end{aligned}$$

Substituting in (15.1)

$$\begin{aligned} ds^2 &= -dx_1^2 - dx_2^2 - dx_3^2 + (1 - \omega^2(x_1^2 + x_2^2))\, dx_4^2 \\ &+ 2\omega\, x_2\, dx_1 dx_4 - 2\omega\, x_1\, dx_2 dx_4. \end{aligned} \tag{15.5}$$

By comparing this with (15.3) we obtain the values of the g's for this system of co-ordinates.

16. These values of the g's express the metrical properties of the space that is being used. But the observer has no immediate perception of them as properties of space. He does not realize that there is anything geometrically unnatural about axes rotating with the earth, but he perceives a field of centrifugal force. Experiments, such as Foucault's pendulum and the gyro-compass, designed to exhibit the absolute rotation of the earth, are more naturally interpreted as detecting this field of force.

Thus the coefficients g_{11} etc., can be taken as specifying a field of force. That they are *sufficient* to define it completely may be seen from the following consideration. The world-line of a particle under no forces is a straight line in the system x, y, z, t, and its equation may be written in the form

$$\int ds \quad \text{is stationary;} \tag{16.1}$$

but in this form the equation is independent of the choice of co-ordinates, and applies to all systems. If we choose new co-ordinates, the world-line given by (16.1) becomes curved and the curvature is attributed to the field of force introduced; but clearly the curvature of the path can only depend on the expression for ds in the new co-ordinates, t.e., on the g's. Thus the force is completely defined by the g"s. It will be noticed that in (15.5)

$$g_{44} = 1 - 2\Omega, \tag{16.2}$$

where $\Omega = \frac{1}{2}\omega^2(x_1^2 + x_2^2)$ = the potential of the centrifugal force.

Thus g_{44} can be regarded as a potential; and by analogy all the coefficients are regarded as components of a generalized potential of the field of force.

According to the principle of equivalence it must also be possible to specify a gravitational field by a set of values of the g's. It will be our object to find the differential equations satisfied by the g's representing a gravitational field. These differential equations for the generalized potential will express the law of gravitation, just as the Newtonian law is expressed by $\nabla^2\varphi = 0$.

The double aspect of these coefficients, g_{11}, etc., should be noted. (1) They express the metrical properties of the co-ordinates. This is the official standpoint of the principle of relativity, which scarcely recognises the term "force." (2) They express the potentials of a field of force. This is the unofficial interpretation which we use when we want to translate our results in terms of more familiar conceptions.

Although we deny absolute space, in the sense that we regard all space-time frameworks in which we can locate natural phenomena as on the same footing, yet we admit that space the whole group of possible spaces – may have some absolute properties. It may, for instance, be homaloidal or non-homaloidal. Whatever the co-ordinates, space near attracting matter is non-homaloidal, space at an infinite distance from matter is homaloidal. You cannot use the same co-ordinates for describing both kinds of space, any more than you can use rectangular co-ordinates on the surface of a sphere; that is, in fact, the geometrical interpretation of the difference. Homaloidal space-time may be regarded as a four-dimensional plane drawn in a continuum of five dimensions; whereas non-homaloidal space-time must be regarded as a curved surface in five dimensions.[3] These considerations apply, of course, to *measured* space; we can always throw the blame on our measuring rods, and apply theoretical corrections to our measures so as to make them agree with any kind of space we please.

It is not necessary, and indeed it is not possible, to draw a sharp distinction between the portions of the g's arising from the choice of co-ordinates and the

[3]We shall see (§44) that in a region, not containing matter, but traversed by a gravitational field due to matter, the Gaussian or total curvature is zero; but such a space-time does not correspond to a plane in five dimensions, or to any surface which can be developed into a plane. The space-time in a gravitational field has an essential curvature in the ordinary sense, although it happens that the particular invariant technically called "the curvature" vanishes. In three-dimensional space a surface with zero Gaussian curvature can always be developed into a plane; but this is not true for space of higher dimensions, so that the three-dimensional analogy is liable to lead to misunderstanding.

portions arising from the gravitation of matter. We have seen that, when there is no field of force, ds^2 has the form (15.1), so that the g's have the values,

$$\begin{matrix} -1 & 0 & 0 & 0 \\ 0 & -1 & 0 & 0 \\ 0 & 0 & -1 & 0 \\ 0 & 0 & 0 & 1 \end{matrix} \tag{16.3}$$

These values then express that there is no field of force, and in the absence of a gravitational field produced by matter it is possible to take our co-ordinates (Galilean co-ordinates) so that the values (16.3) hold everywhere. We naturally regard such co-ordinates as fundamental; and, if we choose any other co-ordinates, the deviations of the g's from this peculiarly simple set of values are regarded as due to the distortion of the space-time chosen. But by §14, when gravitating matter is in the neighborhood, there is no possibility of choosing co-ordinates, so that the values (16.3) hold everywhere, and there is no criterion for selecting any one of the possible systems of co-ordinates as more fundamental than the others.[4]

Accordingly we shall henceforth apply the term "gravitational field" to the whole field of force given by the g's, whatever its origin. In the particular case when no part of it is due to the gravitation of matter, we shall say there is no *permanent* gravitational field.

Just as Galilean co-ordinates are defined by the values (16.3) of the g's, so any other co-ordinates must be defined analytically by specifying the g's as functions of x_1, x_2, x_3, x_4 or – what comes to the same thing – by giving the expression for ds^2. For example, if in two dimensions $ds^2 = dx_1^2 + x_1^2 dx_2^2$ the coordinates as recognized as plane polar co-ordinates with $x_1 = r$, $x_2 = \theta$ if $ds^2 = dx_1^2 + \cos^2 x_1 \, dx_2^2$, the co-ordinates are latitude (x_1) and longitude (x_2) on a sphere. We might take for the 10 g's perfectly arbitrary functions of x_l, x_2, x_3, x_4, and so obtain a ten-fold infinity of mathematically conceivable systems of co-ordinates. But this would include many systems of co-ordinates which describe kinds of space-time not occurring in Nature. In any particular problem our choice is restricted to a four-fold infinity, viz., if x_1, x_2, x_3, x_4 is a possible system, then four arbitrary functions of x_2, x_2, x_3, x_4 will form a possible system. In some other problem there will be an entirely different group of possible systems; the space-times in the two problems have thus certain absolute properties which are irreconcilable, and we interpret this physically by saying that the permanent gravitational field is different in the two cases. Further, taking all possible distributions of permanent gravitational field which can occur in space (in the neighborhood of, but not containing, matter), we do not exhaust the conceivable

[4]Thus if we say "take rectangular axes with the sun as origin" the statement is ambiguous. Unaccepted rectangular axes imply that ds^2 is of the form (15.1) – no other means of defining them having yet been given. Owing to the sun's gravitation there is no system of co-ordinates for which this is true, and several different systems present rival claims to be regarded as the best approximation possible. The difficulty does not arise if we only have to consider an infinitesimal region of space; in that case the co-ordinates (giving "natural measure") are defined without ambiguity.

variety of functions expressing the g's. There is a general limitation on the g's – imposed, not by mathematics, but by Nature which is expressed by the differential equations of the law of gravitation which we are about to seek. The law of gravitation, in fact, expresses certain absolute properties common to all the measured space-times that can under any conditions occur in Nature.

The law of gravitation, or general relation connecting the g's, must hold for all observed values of the g's. Since the g's define the system of co-ordinates used, this means that the relation must hold for all possible systems of co-ordinates. If new co-ordinates are chosen, we find new values of the g's as in (15.5); and the differential equations between the new g's and new co-ordinates must be the same as between the old gs and old co-ordinates. In mathematical language the equations must be covariant.

There is a resemblance between this statement and the statement of §12 which is somewhat deceptive. We there found that observable events have no reference to any particular system of co-ordinates, and therefore all laws of nature can be expressed in a form independent of the co-ordinates. But this alone does not allow us to deduce the covariance of the equations satisfied by the gravitation-potentials. Without the principle of equivalence we could no doubt define the field by certain potentials φ_1, φ_2, $\varphi_3, \dots$ which satisfy differential equations independent of the choice of co-ordinates. But that conveys no information of value, unless we are told how to find $\varphi'_1, \varphi'_2, \dots$ in the co-ordinates x'_1, x'_2, x'_3, x'_4 from the values $\varphi_1, \varphi_2, \dots$ in the co-ordinates x_1, x_2, x_3, x_4 . The statement in §12 tells us nothing about that. It is the principle of equivalence which, by identifying the potentials with the g's for which the method of transformation is known, supplies the missing link.

17. The Newtonian law of gravitation, $\nabla^2 g_{44} = 0$, does not fulfil the condition of covariance nor does any modification of it, which immediately suggests itself. We have, therefore, to seek a new law guided by the condition that it must be expressed by a covariant set of equations between the g's. It will be found in Chapter IV that the choice is so restricted as to leave little doubt as to what the new law must be.

If we write the required equations in the form

$$T_1 = 0, \quad T_2 = 0, \quad T_3 = 0, \quad \text{etc.},$$

the left-hand sides, T_1, T_2, T_3, may be regarded as components of a kind of generalized vector, only the number of components is not, as in a vector, restricted to 4.

The covariance of the equations means that, if all the components vanish in one system, of co-ordinates, they must vanish in all systems. To secure this, $T_1, T_2, \dots$ must obey a linear law of transformation; thus

$$\begin{aligned} T'_1 &= \lambda_1 T_1 + \lambda_2 T_2 + \lambda_3 T_3 + \dots \\ T'_2 &= \mu_1 T_1 + \mu_2 T_2 + \mu_3 T_3 + \dots \end{aligned} \tag{17.1}$$

where the coefficients are functions of the co-ordinates depending on the transformation. Generalized vectors of this kind are called tensors; and it will be necessary for us to study their properties in the next chapter, in order to select the one which can represent the new law of gravitation.

We see that if an equation is known to be a tensor-equation, it is sufficient to prove it for one particular system of co-ordinates; it will then automatically hold in any other system obtainable by a mathematical transformation.

The more general purpose of the tensor theory is this: If we are given a set of equations expressing some physical law in the usual co-ordinates, we may be able to recognize these as the degenerate form for Galilean co-ordinates of some tensor equation. Expressed in tensor form, these equations will then hold for all systems of co-ordinates that can be derived by a mathematical transformation. Subject to the limitations of §14, they will also hold for the gravitational field, although the co-ordinates in that case cannot be obtained by a mathematical transformation. The intermediate step is of no great interest, since the mere transformation of co-ordinates leads to nothing new. But by this mode of approach we obtain the corresponding equations as modified by the action of a gravitational field. This is a very powerful method of investigation.

18. I anticipate that some readers will find the next two chapters difficult, and I therefore place here, out of order, a brief account of the field of a particle according to the new law of gravitation; but doubt if there is any royal road to relativity, and it is scarcely possible to make serious progress except by analytical methods.

We shall find that when a heavy particle is at rest at the origin, the expression for the line-element in plane polar co-ordinates is

$$ds^2 = -\gamma^{-1}dr^2 - r^2d\theta^2 - dz^2 + \gamma dt^2, \tag{18.1}$$

where $\gamma = 1-2m/r$, and m is the mass of the particle, the constant of gravitation and the velocity of light being unity. For the sun, $m = 1,47$ kilometres,[5] so that γ generally differs from unity by a very small quantity.

If $\Omega = m/r =$ Newtonian potential at the point considered, we have

$$g_{44} = \gamma = 1 - 2\Omega \tag{18.2}$$

just as in the case of the centrifugal force (16.2). The Newtonian attraction is therefore a consequence of the coefficient of dt^2.

The general meaning of (18.1) is that our measures will not fit together in Euclidean space. Measuring in the direction r we have

$$i\,ds = \gamma^{-\frac{1}{2}}\,dr,$$

that is to say we must correct the measured length $i\,ds$ in the radial direction, multiplying it by $\gamma^{\frac{1}{2}}$ in order to obtain a length dr which will fit into Euclidean space. Or we may say that our measuring rod contracts when placed radially;

[5]This can be verified roughly as follows: For a circular orbit $m/r^2 = v^2/r$, the constant of gravitation being unity. Applying this to the earth, $v =$ earth's orbital velocity $= 30$ km. per sec. $= 10^{-4}\times$velocity of light. Hence in our units $m = 10^{-8}r$; and r, the radius of the earth's orbit is 1.5×10^8 km.

transverse measures require no correction. Similarly the measured time must multiplied by $\gamma^{-\frac{1}{2}}$, u.e., our clocks run slow.

But there is more than one way of correcting the measures to fit Euclidean space, so that we are not really justified in making precise statements as to the behaviour of our clocks and measuring rods. It is better not to discuss their defects, but to accept the measures and examine the properties of the corresponding non-Euclidean space and time.

If we draw a circle with a heavy particle near the center, the ratio of the measured circumference to the measured diameter will be a little less than π owing to the factor $\gamma^{-\frac{1}{2}}$ affecting radial measures. It is thus like a circle-drawn on a sphere, for which the circumference is less than π times the diameter if we measure along the surface of the sphere. We may imagine space pervaded by a gravitational field to have a curvature in some purely mathematical fifth dimension.

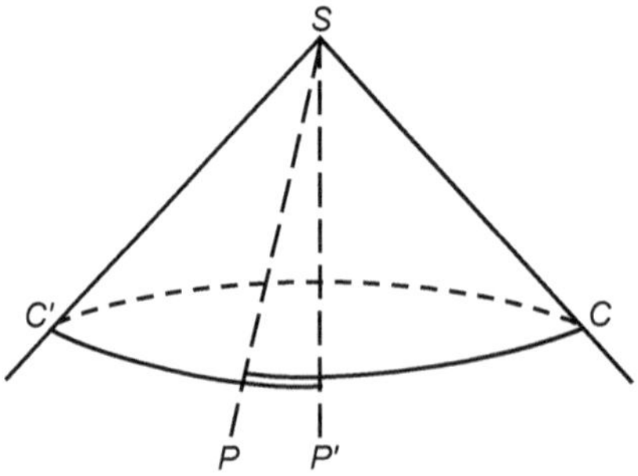

Fig. 3

If we draw the elliptic orbit of a planet, slit it along a radius and try to fold it round our curved space there will evidently be some overlap. For example, take a cone with the sun as apex as roughly representing the curved space. Starting with the radius vector SP, the Euclidean space will fold completely round the cone and overlap to the extent PSP' Thus the corresponding radius advances through an angle PSP' each revolution (Fig. 3). This shows one reason for the advance of perihelion of a planet, which is one of the most important effects predicted by the new theory; but it is not the whole explanation.

The reader may not unnaturally suspect that there is an admixture of metaphysics in a theory which thus reduces the gravitational field to a modification of the metrical properties of space and time. This suspicion, however, is a complete misapprehension, due to the confusion of space, as we have defined it, with some transcendental and philosophical space.

There is nothing metaphysical in the statement that under certain circumstances the measured circumference of a circle is less than π times the measured diameter; it is purely a matter for experiment. We have simply been studying the way in which physical measures of length and time fit together – just as Maxwell's equations describe how electrical and magnetic forces fit together. The trouble is that we have inherited a preconceived idea of the way in which measures, if "true," ought to fit. But the relativity standpoint is that we do not know, and do not care, whether the measures under discussion are "true" or not; and we certainly ought not to be accused of metaphysical speculation, since we confine ourselves to the geometry of measures which are strictly practical, if not strictly practicable. It is desirable to insist that we do not attribute any *causative* properties to these distortions of measured space and time. To hold that a property of our measuring-rods is the cause of gravitation would be as absurd as to hold that the fall of the barometer is the cause of the storm.

3 The Theory of Tensors

19. We consider transformations from one system of co-ordinates x_1, x_2, x_3, x_4 to another system x'_1, x'_2, x'_3, x'_4

(a) Notation
The formula (15.3) for ds^2 may be written

$$ds^2 = \sum_{\mu=1}^{4}\sum_{\nu=1}^{4} g_{\mu\nu}\, dx_\mu\, dx_\nu \quad (g_{\mu\nu} = g_{\nu\mu}). \tag{19.11}$$

In the following work we shall omit the signs of summation, adopting the convention that, whenever a literal suffix appears twice in a term, the term is to be summed for values of the suffix $1, 2, 3, 4$. If a suffix appears once only, no summation is indicated. Thus we shall write (19.11)

$$ds^2 = g_{\mu\nu} dx_\mu\, dx_\nu \tag{19.12}$$

In rare cases it may be necessary to write a term containing a suffix twice which is not to be summed; these cases will always be specially indicated. In general, however, this convention anticipates our desires, and actually gives a kind of momentum in the right direction to the analysis.

As a rule of manipulation it may be noticed that any suffix appearing twice is a dummy, and can be changed freely to any other suffix not occurring in the same term.

(b) Covariant and Contravariant Vectors.
The vector (dx_1, dx_2, dx_3, dx_4) is transformed according to the equations

$$dx'_1 = \frac{\partial x'_1}{\partial x_1} dx_1 + \frac{\partial x'_1}{\partial x_2} dx_2 + \frac{\partial x'_1}{\partial x_3} dx_3 + \frac{\partial x'_1}{\partial x_4} dx_4, \quad \text{etc.,}$$

or, with our convention as to notation

$$dx'_\mu = \frac{\partial x'_\mu}{\partial x_\sigma} dx_\sigma$$

Any vector transformed according to this law is called a *contravariant* vector; its character is denoted by the notation $A^\mu (\mu = 1, 2, 3, 4)$. The law may be written

$$A'^{\mu} = \frac{\partial x'_{\mu}}{\partial x_{\sigma}} A^{\sigma}, \tag{19.21}$$

where, as already explained, summation is indicated by the double appearance of the dummy σ.

If φ is a scalar (i.e., invariant) function of position the vector

$$\left(\frac{\partial \varphi}{\partial x_1}, \frac{\partial \varphi}{\partial x_2}, \frac{\partial \varphi}{\partial x_3}, \frac{\partial \varphi}{\partial x_4}\right)$$

is transformed according to the law

$$\frac{\partial \varphi}{\partial x'_1} = \frac{\partial x_1}{\partial x'_1}\frac{\partial \varphi}{\partial x_1} + \frac{\partial x_2}{\partial x'_1}\frac{\partial \varphi}{\partial x_2} + \frac{\partial x_3}{\partial x'_1}\frac{\partial \varphi}{\partial x_3} + \frac{\partial x_4}{\partial x'_1}\frac{\partial \varphi}{\partial x_4}.$$

A vector transformed according to this law is called a *covariant* vector, denoted by A_{μ} The law may be written

$$A'_{\mu} = \frac{\partial x_{\sigma}}{\partial x_{\mu}} A_{\sigma}. \tag{19-22}$$

A covariant vector is not necessarily the gradient of a scalar.

The customary geometrical conception of a vector does not reveal the distinction between the two classes of contravariant and covariant vectors. We usually represent any directed quantity by a straight line, which should strictly correspond only to a contravariant vector. The other class of directed quantities is more properly represented by the reciprocal of a straight line; but in elementary applications, when we are thinking in terms of rectangular co-ordinates, there is no need to make this distinction. Consider, however, a fluid with a velocity potential. With rectangular co-ordinates the velocity is equal to the gradient of the velocity potential. Both these are directed quantities, i.e., vectors, and the vector relation extends to their rectangular components; thus

$$\frac{dx}{dt} = \frac{\partial \varphi}{\partial x}, \quad \frac{dy}{dt} = \frac{\partial \varphi}{\partial y}, \quad \frac{dz}{dt} = \frac{\partial \varphi}{\partial z}.$$

But if we use oblique axes or curvilinear co-ordinates, the relation no longer holds. E.g., it is not true that in polar co-ordinates $d\theta/dt = \partial\varphi/\partial\theta$; the actual relation is $r\,d\theta/dt = \partial\varphi/r\,d\theta$. This is because the two vectors are of opposite natures, the first being contravariant and the second covariant. If they tad been of the same nature the relation must have held for all systems of co-ordinates. Clearly, since in our work we consider all systems of co-ordinates as on the same footing, we have to distinguish carefully between the two types. We realize at once that the equation $dx_{\mu/dt} = \partial\varphi/\partial x_{\mu}$, being an equation between vectors of opposite kinds, is impossible as a general equation for all systems of co-ordinates, i.e., it is not a covariant equation.

(c) *Tensors of Higher Rank*

We can denote by $A_{\mu\nu}$ a quantity having 16 components, obtained by giving different numerical values to μ and ν. Similarly, $A_{\mu\nu\sigma}$ has 64 components. By a generalization of (19.21) and (19.22) we classify quantities of this kind according to their transformation laws, viz.,

$$\text{Covariant tensors} \qquad A'_{\mu\nu} = \frac{\partial x_\sigma}{\partial x_\mu}\frac{\partial x_\tau}{\partial x'_\nu} A_{\sigma\tau} \tag{19.31}$$

$$\text{Contravariant tensors} \qquad A'^{\mu\nu} = \frac{\partial x'_\mu}{\partial x_\sigma}\frac{\partial x'_\nu}{\partial x_\tau} A^{\sigma\tau} \tag{19.32}$$

$$\text{Mixed tensors} \qquad A'^{\nu}_{\mu} = \frac{\partial x_\sigma}{\partial x'_\mu}\frac{\partial x'_\nu}{\partial x_\tau} A^{\tau}_{\sigma} \tag{19.33}$$

and similarly for tensors of the third and higher rank. These equations of transformation are linear, so that the conditions of §17 are satisfied. Also it is not difficult to see that there can be no other linear types of transformation-laws having the necessary transitive property. For example, consider a vector A_σ. Introducing a third set of co-ordinatesx''_λ, we have

$$A''_\lambda = \frac{\partial x'_\mu}{\partial x''_\lambda} A'_\mu \quad \text{and} \quad A'_\mu = \frac{\partial x_\sigma}{\partial x'_\mu} A_\sigma.$$

But

$$\frac{\partial x_\sigma}{\partial x'_\mu}\frac{\partial x'_\mu}{\partial x''_\lambda} = \frac{\partial x_\sigma}{\partial x''_\lambda}; \quad \text{hence} \quad A''_\lambda = \frac{\partial x_\sigma}{\partial x''_\lambda} A_\sigma,$$

showing that the result is the same whether the transformation is performed in two steps or directly. Other suggested types of transformation law have not this necessary property. Thus all possible types of tensors are included.

Evidently the sum of two tensors of the same character is a tensor.

The product of two tensors is a tensor, and its character is the sum of the characters of the component tensors. For example, consider the product $A_{\mu\nu} B^{\rho}_{\lambda}$, we have by (19.31) and (19.33)

$$A'_{\mu\nu} = \frac{\partial x_\alpha}{\partial x'_\mu}\frac{\partial x_\beta}{\partial x'_\nu} A_{\alpha\beta} \quad B'^{\rho}_{\lambda} = \frac{\partial x_\gamma}{\partial x'_\lambda}\frac{\partial x'_\rho}{\partial x_\delta} B^{\delta}_{\gamma}.$$

Hence

$$(A'_{\mu\nu} B'^{\rho}_{\lambda}) = \frac{\partial x_\alpha}{\partial x'_\mu}\frac{\partial x_\beta}{\partial x'_\nu}\frac{\partial x_\gamma}{\partial x'_\lambda}\frac{\partial x'_\rho}{\partial x_\delta} (A_{\alpha\beta} B^{\delta}_{\gamma}) \tag{19.34}$$

showing that the law of transformation is that of a tensor of the fourth rank having the character denoted by $C^{\rho}_{\mu\nu\lambda}$.

The product of two vectors is a tensor of the second rank, but a tensor of the second rank is not necessarily the product of two vectors.

A familiar example of a tensor of the second rank is afforded by the stresses in a solid or viscous fluid. The component of stress denoted by p_{xy} represents the traction In the y−direction exerted across an interface perpendicular to the x−direction. Each component involves a specification of two directions.

(d) *Inner Multiplication*

If we multiply A_μ by B^μ, the repetition of the suffix involves summation of the resulting products. The result is called the *inner product* in contrast to the ordinary or *outer product* $A_\mu B^\nu$. The notation at once shows whether the product is inner or outer in any formula.

From a mixed tensor such as $A^\tau_{\mu\nu\sigma}$ we can form a *"contracted"* tensor $A^\sigma_{\mu\nu\sigma}$, which is of the second rank with suffixes μ and ν (since σ is now a dummy suffix). To show that it is a tensor we have as in (19.34)

$$A^\sigma_{\mu\nu\sigma} = \frac{\partial x_\alpha}{\partial x'_\mu}\frac{\partial x_\beta}{\partial x'_\nu}\frac{\partial x_\gamma}{\partial x'_\sigma}\frac{\partial x'_\sigma}{\partial x_\delta} A^\delta_{\alpha\beta\gamma}. \tag{19.41}$$

But

$$\frac{\partial x_\gamma}{\partial x'_\sigma}\frac{\partial x'_\sigma}{\partial x_\delta} = \frac{\partial x_\gamma}{\partial x_\delta} = 0 \text{ or } 1, \text{ according as } \gamma \neq \delta \text{ or } \gamma = \delta.$$

Hence

$$\frac{\partial x_\gamma}{\partial x'_\sigma}\frac{\partial x'_\sigma}{\partial x_\delta} A^\delta_{\alpha\beta\gamma} = 0 + 0 + 0 + A^\gamma_{\alpha\beta\gamma}.$$

Substituting in (19.41) we see that $A^\sigma_{\mu\nu\sigma}$ follows the law of transformation (19.31) and is therefore a covariant tensor.

An expression such as $A^\tau_{\mu\sigma\sigma}$ is not a tensor, and no interest attaches to it.

By a similar argument we see that A^μ_μ, $A^{\mu\nu}_{\mu\nu}$ are invariant, and consequently $A_\mu B^\mu$ is an invariant. An invariant, or scalar, corresponds to a tensor of zero rank.

(e) *Criterion for the Tensor Character*

To prove that a given quantity is a tensor, we either find directly its equations of transformation, or we express it as the sum or product of other tensors, or, under certain restrictions, as the quotient of two tensors according to the following theorem: A quantity, which on inner multiplication by *any* covariant (alternatively, by *any* contravariant) vector always gives a tensor, is itself a tensor.

To prove this, suppose that $A_{\mu\nu}B^\nu$ is a covariant vector for any choice of the contravariant vector B^ν . Then by (19.22)

$$A'_{\mu\nu}B'^\nu = \frac{\partial x_\sigma}{\partial x'_\mu} A_{\sigma\tau}B^\tau.$$

But by (19.21) applied to the inverse transformation from accented to unaccented letters,

$$B^\tau = \frac{\partial x_\tau}{\partial x'_\nu} B'^\nu.$$

Hence

$$B^\nu \left(A'_{\mu\nu} - \frac{\partial x_\sigma}{\partial x'_\mu}\frac{\partial x_\tau}{\partial x'_\nu} A_{\sigma\tau} \right) = 0.$$

Since B'^{ν} is arbitrary, the quantity in the bracket must vanish, showing that $A_{\mu\nu}$ is a covariant tensor (19.31). The proof can evidently be extended to tensors of any character.

20. *(a) The Fundamental Tensor*

Since $g_{\mu\nu}dx_{\mu}dx_{\nu} = ds^2$, which is an invariant or tensor of zero order, and dx_{ν} is an arbitrary contravariant vector, it follows from the last theorem that $g_{\mu\nu}dx_{\mu}$ is a covariant tensor of the first rank. Repeating the argument, since dx_{μ} is an arbitrary contravariant vector, $g_{\mu\nu}$ must be a covariant tensor of the second rank.

The determinant formed with the elements $g_{\mu\nu}$ is called the fundamental determinant and is denoted by g.

We define $g^{\mu\nu}$ to be the minor of $g_{\mu\nu}$ divided by g.

From this definition $g_{\mu\sigma}g^{\mu\nu}$ reproduces the fundamental determinant divided by itself, when $\sigma = \nu$, and gives a determinant with two rows identical, when $\sigma \neq \nu$. We write

$$\begin{aligned} g^{\nu}_{\sigma} = g_{\mu\sigma}g^{\mu\nu} &= 1 \text{ when } \sigma = \nu \\ &= 0 \text{ when } \sigma \neq \nu. \end{aligned} \tag{20.1}$$

Hence if A^{ν} is an arbitrary contravariant vector

$$g^{\nu}_{\sigma}A^{\sigma} = A^{\nu} + 0 + 0 + 0 = A^{\nu}. \tag{20.15}$$

This shows by the theorem of §19 *(e)* that g^{ν}_{σ} is a tensor, and it evidently is a mixed tensor as the notation has anticipated.[1]

Similarly, since $g_{\mu\sigma}A^{\sigma}$ a is a covariant vector, arbitrary on account of the free choice of A^{σ}, and $g^{\mu\nu}g_{\mu\sigma}\, A^{\sigma} = A_{\nu}$, $g^{\mu\nu}$ must be a contravariant tensor.

We Lave thus the three fundamental tensors

$$g_{\mu\nu} \quad g^{\nu}_{\mu}, \quad \text{and} \quad g^{\mu\nu},$$

of covariant, mixed and contravariant characters.

It will be seen from (20.15) that g^{ν}_{σ} acts as a *substitution operator* – substituting ν for σ in the operand.

vspace3mm

(b) Associated Tensors

With any covariant tensor $A_{\mu\nu}$ we can associate

$$\text{a mixed tensor} \quad A^{\nu}_{\mu} = g^{\nu\alpha}A_{\mu\alpha} \tag{20.21}$$

$$\text{a contravariant tensor} \quad A^{\mu\nu} = g^{\mu\alpha}g^{\nu\beta}\, A_{\alpha\beta} = g^{\mu\alpha}\, A^{\nu}_{\alpha} \tag{20.22}$$

$$\text{a scalar} \quad A = g^{\mu\nu}A_{\mu\nu} = A^{\mu}_{\mu} \tag{20.23}$$

[1] In applying the theorem of §19 *(e)*, the appropriate notation for the tensor (expressing its covariant or contravariant character) is found by inspection. An equation such as (20.15) must have the suffixes on both sides In corresponding positions; the upper and lower σ on the left cancel one another. Cf. equations (20.21), (20.22), (20.23). It must be noted, however, that in an expression such as $g_{\mu\nu}dx_{\mu}$, dx_{μ} is contravariant, so that the second μ is really an upper suffix.

(c) The Jacobian

Denoting the determinant formed with elements $a_{\mu\nu}$ by $|a_{\mu\nu}|$, the Jacobian of the transformation is

$$J = \frac{\partial(x'_1 x'_2 x'_3 x'_4)}{\partial(x_1 x_2 x_3 x_4)} = \left|\frac{\partial x'_\mu}{\partial x_\sigma}\right|$$

Now

$$g' = |g'_{\mu\nu}| = \left|\frac{\partial x_\sigma}{\partial x'_\mu}\frac{\partial x_\tau}{\partial x'_\nu} g_{\sigma\tau}\right| \quad \text{by (19.31)}$$

$$= \left|\frac{\partial x_\sigma}{\partial x'_\mu}\right| \times \left|\frac{\partial x_\tau}{\partial x'_\nu}\right| \times |g_{\sigma\tau}| \quad \text{not summed,}$$

since in our notation the ordinary rule for multiplying determinants is $|A_{\alpha\beta}| \times |B_{\alpha\gamma}| = |A_{\alpha\beta}B_{\alpha\gamma}|$ (left side not summed).

Hence

$$g' = \frac{1}{J^2} g.$$

If $d\tau$ is an element of four-dimensional volume, we have

$$d\tau' = J\, d\tau,$$

so that

$$\sqrt{-g}\,.\,d\tau = \sqrt{-g'}\,.\,d\tau'. \tag{20.3}$$

We shall always assume that the Jacobian is finite, i.e., that the transformation has no singularity in the region considered. The determinant g is always negative for real transformations.

21. *Auxiliary Formula for the Second Derivatives*

We introduce certain quantities known as Christoffel's 3-index symbols, viz.,

$$[\mu\nu, \lambda] = \frac{1}{2}\left(\frac{\partial g_{\mu\lambda}}{\partial x_\nu} + \frac{\partial g_{\nu\lambda}}{\partial x_\mu} - \frac{\partial g_{\mu\nu}}{\partial x_\lambda}\right) \tag{21.11}$$

$$\{\mu\nu, \lambda\} = \frac{1}{2} g^{\lambda\alpha}\left(\frac{\partial g_{\mu\alpha}}{\partial x_\nu} + \frac{\partial g_{\nu\alpha}}{\partial x_\mu} - \frac{\partial g_{\mu\nu}}{\partial x_\alpha}\right). \tag{21.12}$$

We have

$$\{\mu\nu, \lambda\} = g^{\lambda\alpha}\,[\mu\nu, \alpha] \tag{21.13}$$

and the reciprocal relation follows by (20.1)

$$[\mu\nu, \lambda] = g_{\lambda\alpha}\,\{\mu\nu, a\}. \tag{21.14}$$

Since $g_{\mu\nu}$ is a covariant tensor

$$g'_{\mu\nu} = \frac{\partial x_\alpha}{\partial x'_\mu}\frac{\partial x_\beta}{\partial x'_\nu} g_{\alpha\beta}.$$

Hence

$$\frac{\partial g'_{\mu\nu}}{\partial x'_\lambda} = g_{\alpha\beta}\left\{\frac{\partial^2 x_\alpha}{\partial x'_\mu \partial x'_\lambda}\frac{\partial x_\beta}{\partial x'_\nu} + \frac{\partial^2 x_\alpha}{\partial x'_\nu \partial x'_\lambda}\frac{\partial x_\beta}{\partial x'_\mu}\right\} + \frac{\partial x_\alpha}{\partial x'_\mu}\frac{\partial x_\beta}{\partial x'_\nu}\frac{\partial x_\gamma}{\partial x'_\lambda}\frac{\partial g_{\alpha\beta}}{\partial x_\gamma}. \tag{21.15}$$

In the second term in the bracket we have interchanged α and β, which is legitimate since they are dummies; in the last term we have used

$$\frac{\partial}{\partial x'_\lambda} = \frac{\partial x_\gamma}{\partial x'_\lambda}\frac{\partial}{\partial x_\gamma}.$$

Similarly,

$$\frac{\partial g'_{\nu\lambda}}{\partial x'_\mu} = g_{\alpha\beta}\left\{\frac{\partial^2 x_\alpha}{\partial x'_\nu \partial x'_\mu}\frac{\partial x_\beta}{\partial x'_\lambda} + \frac{\partial^2 x_\alpha}{\partial x'_\lambda \partial x'_\mu}\frac{\partial x_\beta}{\partial x'_\nu}\right\} + \frac{\partial x_\alpha}{\partial x'_\mu}\frac{\partial x_\beta}{\partial x'_\nu}\frac{\partial x_\gamma}{\partial x'_\lambda}\frac{\partial g_{\beta\gamma}}{\partial x_\alpha}. \tag{21.16}$$

$$\frac{\partial g'_{\mu\lambda}}{\partial x'_\nu} = g_{\alpha\beta}\left\{\frac{\partial^2 x_\alpha}{\partial x'_\mu \partial x'_\nu}\frac{\partial x_\beta}{\partial x'_\lambda} + \frac{\partial^2 x_\alpha}{\partial x'_\lambda \partial x'_\nu}\frac{\partial x_\beta}{\partial x'_\mu}\right\} + \frac{\partial x_\alpha}{\partial x'_\mu}\frac{\partial x_\beta}{\partial x'_\nu}\frac{\partial x_\gamma}{\partial x'_\lambda}\frac{\partial g_{\alpha\gamma}}{\partial x_\beta}. \tag{21.17}$$

where in the last term we have made some interchanges of the dummy suffixes α, β, γ.

Adding these two equations and subtracting (21.15) we have by (21.11)

$$[\mu\nu, \lambda]' = g_{\alpha\beta}\frac{\partial^2 x_\alpha}{\partial x'_\mu \partial x'_\nu}\frac{\partial x_\beta}{\partial x'_\lambda} + \frac{\partial x_\alpha}{\partial x'_\mu}\frac{\partial x_\beta}{\partial x'_\nu}\frac{\partial x_\gamma}{\partial x'_\lambda}[\alpha\beta, \gamma]. \tag{21.18}$$

Multiply through by $g'^{\lambda\rho}\,\frac{\partial x_{epsilon}}{\partial x'_\rho}$ we have

$$\begin{aligned}
\{\mu\nu, \rho\}'\frac{\partial x_\varepsilon}{\partial x'_\rho} &= \left(g'_{\lambda\rho}\frac{\partial x_\beta}{\partial x'_\lambda}\frac{\partial x_\varepsilon}{\partial x'_\rho}\right) g_{\alpha\beta}\frac{\partial^2 x_\alpha}{\partial x'_\mu \partial x'_\nu} + \left(g'_{\lambda\rho}\frac{\partial x_\gamma}{\partial x'_\lambda}\frac{\partial x_\varepsilon}{\partial x'_\rho}\right)\frac{\partial x_\alpha}{\partial x'_\mu}\frac{\partial x_\beta}{\partial x'_\nu}[\alpha\beta, \gamma] \\
&= g^{\beta\varepsilon} g_{\alpha\beta}\frac{\partial^2 x_\alpha}{\partial x'_\mu \partial x'_\nu} + \frac{\partial x_\alpha}{\partial x'_\mu}\frac{\partial x_\beta}{\partial x'_\nu}[\alpha\beta, \gamma] \quad \text{by (19.32)} \\
&= \frac{\partial^2 x_\varepsilon}{\partial x'_\mu \partial x'_\nu} + \frac{\partial x_\alpha}{\partial x'_\mu}\frac{\partial x_\beta}{\partial x'_\nu}\{\alpha\beta, \varepsilon\}
\end{aligned} \tag{21.2}$$

using (20.1) and (21.13).

This somewhat complicated formula for $\frac{\partial^2 x_\varepsilon}{\partial x'{\mu}\partial x'_\nu}$ in terms of the first derivatives is needed for the developments in the next paragraph.

22. Covariant Differentiation

If we differentiate a scalar quantity we obtain a tensor (a covariant vector); but if we differentiate a tensor of the first or higher rank the result is not a tensor. We can, however, obtain a tensor which plays the part of a derivative by a more general process. The process is particularly useful in generalizing results which have been obtained in Galilean co-ordinates, since the simple derivative is the degenerate form for Galilean co-ordinates of the covariant derivative here considered.

If A_μ is a covariant vector, then by (19.22)

$$A'_\mu = \frac{\partial x_\sigma}{\partial x'_\mu} A_\sigma.$$

Whence, differentiating,

$$\frac{\partial A'_\mu}{\partial x'_\nu} = \frac{\partial x_\sigma}{\partial x'_\mu}\frac{\partial x_\tau}{\partial x'_\nu}\frac{\partial A_\sigma}{\partial x_\tau} + \frac{\partial^2 x_\sigma}{\partial x'_\mu \partial x'_\nu} A_\sigma.$$

Substitute for $\frac{\partial^2 x_\sigma}{\partial x'_\mu \partial x'_\nu}$ by (21.2); we have

$$\frac{\partial A'_\mu}{\partial x'_\nu} - \{\mu\nu, \rho\}' A_\sigma \frac{\partial x_\sigma}{\partial x'_\rho} = \frac{\partial x_\sigma}{\partial x'_\mu}\frac{\partial x_\tau}{\partial x'_\nu}\frac{\partial A_\sigma}{\partial x_\tau} - \frac{\partial x_\alpha}{\partial x'_\mu}\frac{\partial x_\beta}{\partial x'_\nu}\{\alpha, \sigma\} A_\sigma. \tag{22.1}$$

But $A_\sigma \frac{\partial x_\sigma}{\partial x'_\rho} = A'_\rho$ by (19.22); and in the last term the dummies α, β, σ may be replaced by σ, τ, ρ. Hence if we write

$$A_{\mu\nu} = \frac{\partial A_\mu}{\partial x_\nu} - \{\mu\nu, \rho\} A_\rho \tag{22.2}$$

we have

$$A'_{\mu\nu} = \frac{\partial x_\sigma}{\partial x'_\mu}\frac{\partial x_\tau}{\partial x_\nu} A_{\sigma\tau},$$

showing that $A_{\mu\nu}$ is a tensor. This is called the *covariant derivative* of A_μ.

If A_λ, B_μ are covariant vectors, $A_{\lambda\nu}$, $B_{\mu\nu}$ their covariant derivatives, then

$$A_{\lambda\nu}B_\mu + A_\lambda B_{\mu\nu}$$

is the sum of two tensors, and is therefore a tensor. Substituting from (22.2) this tensor becomes

$$\frac{\partial(A_\lambda B_\mu)}{\partial x_\nu} - \{\lambda\nu, \varepsilon\} A_\varepsilon B_\mu - \{\mu\nu, \varepsilon\} A_\lambda B_\varepsilon \tag{22.3}$$

which is called the derivative of the tensor A_λ, B_μ. It is not difficult to show that any tensor of the second rank can be expressed as the sum of products of four pairs of vectors, and hence (22.3) can be generalized, giving for the covariant derivative of $A_{\lambda\mu}$

$$A_{\lambda\mu\nu} = \frac{\partial A_{\lambda\mu}}{\partial x_\nu} - \{\lambda\nu, \varepsilon\} A_{\varepsilon\mu} - \{\mu\nu, \varepsilon\} A_{\lambda\varepsilon}. \tag{22.4}$$

In a somewhat similar manner formulae for the covariant derivatives of contravariant and mixed tensors can be obtained, viz.,

$$A^\mu_\nu = \frac{\partial A^\mu}{\partial x_\nu} + \{\nu\varepsilon, \mu\} A^\varepsilon. \tag{22.5}$$

$$A^{\lambda\mu}_\nu = \frac{\partial A^{\lambda\mu}}{\partial x_\nu} + \{\nu\varepsilon, \lambda\} A^{\varepsilon\mu} + \{\nu\varepsilon, \mu\} A^{\lambda\varepsilon}. \tag{22.6}$$

$$A^\mu_{\lambda\nu} = \frac{\partial A^\mu_\lambda}{\partial x_\nu} - \{\nu\lambda, \varepsilon\} A^\mu_\varepsilon + \{\nu\varepsilon, \mu\} A^\varepsilon_\lambda. \tag{22.7}$$

The unsymmetrical behaviour of covariant and contravariant indices in these formula should be noticed. In all cases differentiation adds one unit of covariant character.

When the g's s have Galilean values (or, more generally, are constants) the Christoffel symbols vanish, and these derivatives reduce in all cases to the ordinary differential coefficients.

23. The Riemann-Christoffel Tensor

Let us form the second covariant derivative of the vector A_μ that is to say in formula (22.4) we give the tensor $A_{\lambda\mu}$ the value (22.2).

$$
\begin{aligned}
A_{\mu\nu\sigma} &= \frac{\partial}{\partial x_\sigma}\left\{\frac{\partial A_\mu}{\partial x_\nu} - \{\mu\nu,\rho\}\,A_\rho\right\} - \{\mu\sigma,\varepsilon\},\left\{\frac{\partial A_\varepsilon}{\partial x_\nu} - \{\varepsilon\nu,\rho\}\,A_\rho\right\}\\
&\quad - \{\nu\sigma,\varepsilon\}\left\{\frac{\partial A_\mu}{\partial x_\varepsilon} - \{\mu\varepsilon,\rho\}\,A_\rho\right\}\\
&= \frac{\partial^2 A_\mu}{\partial x_\sigma \partial x_\nu} - \{\mu\nu,\rho\}\,\frac{\partial A_\rho}{\partial x_\sigma} - -\{\mu\sigma,\varepsilon\}\,\frac{\partial A_\varepsilon}{\partial x_\nu} - \{\nu\sigma,\varepsilon\}\,\frac{\partial A_\mu}{\partial x_\varepsilon}\\
&\quad + \{\nu\sigma,\varepsilon\}\,\{\mu\varepsilon,\rho\}\,A_\rho + \{\mu\sigma,\varepsilon\}\,\{\varepsilon\nu,\rho\}\,A_\rho - A_\rho\,\frac{\partial}{\partial x_\sigma}\,\{\mu\nu,\rho\}.
\end{aligned}
$$

The first five terms are unaltered by interchanging ν and σ, i.e, by changing the order of differentiation. (We can write ε for ρ in the second term.) Hence

$$A_{\mu\nu\sigma} - A_{\mu\sigma\nu} =$$

$$\left[\{\mu\sigma,\varepsilon\}\,\{\varepsilon\nu,\rho\} - \{\mu\nu,\varepsilon\}\,\{\varepsilon\sigma,\rho\} + \frac{\partial}{\partial x_\nu}\,\{\mu\sigma,\rho\} - \frac{\partial}{\partial x_\sigma}\,\{\mu\nu,\rho\}\right] A_\rho.$$

The left side is a tensor, and A_ρ is an arbitrary covariant vector; therefore, by §19 *(e)* the quantity in the bracket is a tensor. This is called the Riemann-Christoffel tensor, and is denoted by

$$B^\rho_{\mu\nu\sigma} = \{\mu\sigma,\varepsilon\}\,\{\varepsilon\nu,\rho\} - \{\mu\nu\varepsilon\}\,\{\varepsilon\sigma,\rho\} + \frac{\partial}{\partial x_\nu}\{\mu\sigma,\rho\} - \frac{\partial}{\partial x_\sigma}\{\mu\nu,\rho\}. \qquad (23)$$

24. Conditions for Vanishing of the Riemann-Christoffel Tensor

From the foregoing definition the primary meaning of the vanishing of this tensor is that the order of differentiation is indifferent (as in the ordinary differentiation). But the tensor has an even more important property. It will be seen on inspection that it vanishes when the g's have their constant Galilean values.[2] But, since it is a tensor, it must also vanish in any other system of co-ordinates derivable by a mathematical transformation. Thus the equation

$$B^\rho_{\mu\nu\sigma} = 0 \qquad (24.1)$$

[2]The Christoffel symbols vanish when the g's are constants.

is a necessary condition that with suitable choice of co-ordinates ds^2 can be reduced to the form

$$ds^2 = -dx^2 - dy^2 - dz^2 + dt^2. \tag{24.2}$$

In other words it is a necessary condition for the absence of a permanent gravitational field.

It can be shown that the condition is also sufficient. Equation (24.1) contains 96 apparently different equations, since, owing to the antisymmetry in σ and ν, there are only 6 combinations of σ and ν to be combined with 16 combinations of μ and ρ. But these are not all independent, and the number can be reduced to 20, which can be shown to be the number of conditions required for the transformation to the form (24.2) to be possible.

The reduction is effected by writing

$$(\mu\tau\sigma\nu) = g_{\tau\rho}B^{\rho}_{\mu\nu\sigma},$$

so that

$$B^{\rho}_{\mu\nu\sigma} = g^{\lambda\rho}(\mu\lambda\sigma\nu) \quad \text{by (20.1).}$$

Equation (24.1) is thus equivalent to

$$(\mu\tau\sigma\nu) = 0$$

and *vice versa.*

On working out the value of $(\mu\tau\sigma\nu)$ it is seen by inspection that the following additional relations exist :

$$(\mu\tau\sigma\nu) \equiv -(\tau\mu\sigma\nu) \equiv (\nu\sigma\tau\mu) \equiv (\sigma\nu\mu\tau)$$

$$(\mu\tau\sigma\nu) + (\mu\sigma\nu\tau) + (\mu\nu\tau\sigma) \equiv 0,$$

which reduce the number of independent conditions to 20.

25. To sum up what has been accomplished in this chapter, we have discussed the theory of tensors – expressions which have the property that a linear relation between tensors of the same character will hold in all systems of co-ordinates if it holds in one system. We have shown that the tensor-property can be established either by determining the law of transformation, or exhibiting the quantity as a sum or product of other tensors, or, under certain restrictions, as the quotient of tensors. We have found formulae for tensors which play the part of derivatives. Finally, we have found the necessary and sufficient relation between the $g_{\mu\nu}$ which must be satisfied in all systems of co-ordinates, when there is no permanent gravitational field.

This last result is an important step towards obtaining the law of gravitation. Any set of values of the g's which satisfy (24.1) will correspond to a *possible set* of co-ordinates which can be used for describing space not containing a permanent gravitational field. Hence if (24.1) is satisfied the g's are such as can occur in Nature, and are accordingly not inconsistent with the law of gravitation. The required equations of the law of gravitation must, therefore, include the vanishing of the Riemann-Christoffel tensor as a special case.

4 EINSTEIN'S LAW OF GRAVITATION

26. We have seen in §16 that the law of gravitation must be expressed as a set of differential equations satisfied by the g's. We have further found the equations (24.1) which are satisfied in the absence of (i.e., at an infinite distance from) attracting matter. Clearly the general equations between the g's must be covariant equations automatically satisfied when (24.1) is satisfied; but they must be less stringent, so as to admit of permanent gravitational fields, which, we know, do not satisfy (24.1).

The simplest set of equations that suggests itself is

$$G_{\mu\nu} \equiv B^{\rho}_{\mu\nu\rho} = O \tag{26.1}$$

$G_{\mu\nu}$ being the contracted Riemaiin-Christoffel tensor, formed by setting $\sigma = \rho$ and summing. It is evidently satisfied when all components of the Riemann-Christoffel tensor vanish; and it is a less stringent condition.

The equations $G_{\mu\nu} = 0$ are taken by Einstein for the Law of Gravitation. Written in full they are, by (23)

$$-\frac{\partial}{\partial x_\rho}\{\mu\nu,\rho\} + \{\mu\rho,\varepsilon\}\{\nu\varepsilon,\rho\} + \frac{\partial}{\partial x_\nu}\{\mu\rho,\rho\} - \{\mu\nu,\varepsilon\}\{\varepsilon\rho,\rho\} = 0 \tag{26.2}$$

The last two terms can be simplified. We have

$$\{\mu\rho,\rho\} = \frac{1}{2}g^{\rho\epsilon}\left(\frac{\partial g_{\mu\varepsilon}}{\partial x_\rho} + \frac{\partial g_{\rho\varepsilon}}{\partial x_\mu} - \frac{\partial g_{\mu\rho}}{\partial x_\varepsilon}\right) = \frac{1}{2}g^{\rho\varepsilon}\frac{\partial g_{\rho\varepsilon}}{\partial x_\mu}$$

the other terms cancelling on summation.

Hence, since $g^{\rho\varepsilon}g$ is the minor of the element $g_{\rho\varepsilon}$ in the determinant g,

$$\{\mu\rho,\rho\} = \frac{1}{2g}\frac{\partial g}{\partial x_\mu} = \frac{\partial}{\partial x_\mu}\log\sqrt{-g}. \tag{26.25}$$

Equation (26.2) thus becomes

$$\begin{aligned} G_{\mu\nu} = &-\frac{\partial}{\partial x_\rho}\{\mu\nu,\rho\} + \{\mu\rho,\varepsilon\}\{\nu\varepsilon,\rho\} + \frac{\partial^2}{\partial x_\mu \partial x_\nu}\log\sqrt{-g} \\ &- \{\mu\nu,\varepsilon\}\frac{\partial}{\partial x_\epsilon}\log\sqrt{-g} = 0. \end{aligned} \tag{26.3}$$

The equation is symmetrical in μ and ν, and therefore represents 10 different equations. Actually there exist four identical relations between these, so that the number of independent equations is reduced to six (see §39).

The selection of this law of gravitation is not so arbitrary as it might appear. There is no other set of equations corresponding to a tensor of the second rank containing only first and second derivatives of the $g_{\mu\nu}$ and linear in the second derivatives. Moreover, there is no other way of building up a tensor of lower rank out of the components of $B^{\rho}_{\mu\nu\sigma}$.[1]

Having regard to the summations involved in (26.3) it will be seen that the application of the new law of gravitation must involve a considerable amount of calculation. There are first to be calculated 40 different Christoffel symbols, each of which is the sum of 12 terms. Then each of the 10 equations contains 25 terms – chiefly products or derivatives of the Christoffel symbols. Finally the partial differential equations have to be solved. It will probably be admitted that it is worth while to find out whether this suggested law of gravitation will agree with observation before resorting to something more complicated.

27. We are now in a position to define the Principle of Equivalence more precisely. The difference between a permanent gravitational field and an artificial one arising from a transformation of Galilean co-ordinates is that in the latter case (24.1) is satisfied, whereas in the former the less stringent condition (26.1) is satisfied. These equations determine the second differential coefficients of the $g_{\mu\nu}$, so that we can make the natural and artificial fields correspond as far as first differential coefficients, but not in the second differential coefficients. We shall therefore state the Principle of Equivalence as follows:

All laws, relating to phenomena in a geometrical field of force, *which depend on the g's and their first derivatives*, will also hold in a permanent gravitational field. Laws which depend on the second derivatives of the g's will not necessarily apply.

It must be remembered that we give no proof of this; it is merely an explicit statement of our assumptions. It would be quite consistent with the general idea of relativity if the true expression of such laws involved the Riemann-Christoffel tensor, which vanishes in the artificial field, and would have to be replaced before the equations were applied to the gravitational field. But if we were to admit that, the principle of equivalence would become absolutely useless.

The Gravitational Field of a Particle

28. We have seen that the gravitational-potentials satisfy the equations (26.3)

$$\begin{aligned} G_{\sigma\tau} \equiv -\frac{\partial}{\partial x_\alpha}\{\sigma\tau,\alpha\} + \{\sigma\alpha,\beta\}\{\tau\beta,\alpha\} + \frac{\partial^2}{\partial x_\sigma \partial x_\tau}\log\sqrt{-g} \\ -\{\sigma\tau,\alpha\}\frac{\partial}{\partial x_\alpha}\log\sqrt{-g} = 0. \end{aligned} \tag{28.1}$$

[1]The tensor $B^{\rho}_{\rho\nu\sigma}$ vanishes identically. Other suggestions such as $g^{\mu\sigma}\, B^{\rho}_{\mu\nu\sigma}$ merely give a set of equations equivalent to (26.1). The single equation $g^{\mu\nu}\, G_{\mu\nu} = 0$ would, obviously be insufficient to determine the gravitational field.

We shall now find a solution of these equations corresponding to the field of a particle at rest at the origin of space-co- ordinates. We choose polar co-ordinates, viz,

$$x_1 = r, \quad x_2 = \theta, \quad x_3 = \varphi, \quad x_4 = t.$$

In making this statement we are departing somewhat from the standpoint of general relativity. Strictly speaking, we can only define a system of co-ordinates by the form of the corresponding expression for ds^2, that is by the gravitational potentials. So that to specify the co-ordinates that are used involves solving the problem. Further, we have at present no knowledge of a particle of matter, except that it must be a point where the equations (28.1), which hold at points outside matter, break down; we can only distinguish a particle from other mathematically possible singularities, such as doublets, by the symmetry of the resulting field. Thus the logical course is to find a solution, and afterwards discuss what distribution of matter and what system of co-ordinates it represents. We shall, however, find it more profitable to accept the guidance of our current approximate ideas in order to arrive at the required solution inductively.

The line-element ds can be assumed to be of the form

$$ds^2 = -e^\lambda\, dr^2 - e^\mu (r^2 d\theta^2 + r^2 \sin^2\theta\, d\varphi^2) + e^\nu\, dt^2. \tag{28.21}$$

where λ, μ, ν are functions of r only.

The omission of the product terms, $dr\, d\theta$, $dr\, d\varphi$, $d\theta\, d\varphi$, is justified by the symmetry of polar co-ordinates; the omission of $drdt, d\theta dt, d\varphi dt$ involves the symmetry of a static field with respect to past and future time. If the latter products were present we should interpret the co-ordinates as changing with the time.

A further simplification can be made by writing $r^2\, e^\mu = r'^2$ and adopting r' as our new co-ordinate (dropping the accent). The resulting change in dr^2 is absorbed by taking a new λ. Thus the coefficient e^μ is made to disappear and we have

$$ds^2 = -e^\lambda\, dr^2 - r^2 d\theta^2 - r^2 \sin^2\theta\, d\varphi^2 + e^\nu\, dt^2. \tag{28.22}$$

Comparing (28.22) with (15.3), we have

$$g_{11} = -e^\lambda, \quad g_{22} = -r^2, \quad g_{33} = -r^2 \sin^2\theta, \quad g_{44} = e^\nu \tag{28.31}$$

$$\text{and} \qquad g_{\sigma\tau} = 0 \text{ when } \sigma \neq \tau.$$

The determinant g reduces to its leading diagonal, so that

$$-g = e^{\lambda+\nu} r^4 \sin^2\theta, \tag{28.32}$$

$$\text{and} \quad g^{\sigma\sigma} = \frac{1}{g_{\sigma\sigma}}. \tag{28.33}$$

We can now calculate the three-index symbols (21.12)

$$\{\sigma\tau, \alpha\} = \frac{1}{2}\, g^{\alpha\beta} \left\{ \frac{\partial g_{\beta\sigma}}{\partial x_\tau} + \frac{\partial g_{\beta\tau}}{\partial x_\sigma} - \frac{\partial g_{\sigma\tau}}{\partial x_\beta} \right\}.$$

Since the g's vanish except when the two suffixes agree, the summation disappears and we have

$$\{\sigma\tau,\alpha\} = \frac{1}{2g_{\alpha\alpha}}\left\{\frac{\partial g_{\alpha\sigma}}{\partial x_\tau} + \frac{\partial g_{\alpha\tau}}{\partial x_\sigma} - \frac{\partial g_{\sigma\tau}}{\partial x_\alpha}\right\} \quad \text{not summed.}$$

If σ, τ, ρ are unequal we get the following possible cases:

$$\begin{aligned}
\{\sigma\sigma,\sigma\} &= \frac{1}{2}\frac{\partial}{\partial x_\sigma}\log g_{\sigma\sigma} \\
\{\sigma\sigma,\tau\} &= -\frac{1}{2g_{\tau\tau}}\frac{\partial}{\partial x_\tau}g_{\sigma\sigma} \\
\{\sigma\tau,\tau\} &= \frac{1}{2}\frac{\partial}{\partial x_\sigma}\log g_{\tau\tau} \\
\{\sigma\tau,\rho\} &= 0.
\end{aligned} \tag{28.4}$$

None of the above expressions are to be summed.

Whence by (28.31), denoting differentiation with respect to r by accents, we obtain

$$\begin{aligned}
\{11,1\} &= \frac{1}{2}\lambda' \\
\{12,2\} &= \frac{1}{r} \\
\{13,3\} &= \frac{1}{r} \\
\{14.4\} &= \frac{1}{2}\nu' \\
\{22,1\} &= -re^{-\lambda} \\
\{23,3\} &= \cot\theta \\
\{33,1\} &= -r\,\sin^2\theta\, e^{-\lambda} \\
\{33,2\} &= -\sin\theta\,\cos\theta \\
\{44,1\} &= \frac{1}{2}e^{\nu-\lambda}\,\nu'
\end{aligned} \tag{28.5}$$

The remaining 31 Christoffel symbols are zero. It should be noted that $\{21,2\}$ is the same as $\{12,2\}$, etc.

It is now not difficult to obtain the equations of the field. To assist the reader in carrying through the substitutions, we shall write out in full the equations (28.1) omitting the terms (223 in number), which obviously vanish. The following come respectively from G_{11}, G_{22}, G_{33}, $G_{44} = 0$:

$$-\frac{\partial}{\partial r}\{11,1\} + \{11,1\}\{11,1\} + \{12,2\}\{12,2\} + \{13,3\}\{13,3\} + \{14,4\}\{14,4\} + \frac{\partial^2}{\partial r^2}\log\sqrt{-g} - \{11,1\}\frac{\partial}{\partial r}\log\sqrt{-g} = 0$$

$$-\frac{\partial}{\partial r}\{22,1\} + 2\{22,1\}\{12,2\} + \{23,3\}\{23,3\} + \frac{\partial^2}{\partial \theta^2}\log\sqrt{-g} - \{22,1\}\frac{\partial}{\partial r}\log\sqrt{-g} = 0$$

$$-\frac{\partial}{\partial r}\{33,1\} - \frac{\partial}{\partial \theta}\{33,2\} + 2\{33,1\}\{13,3\} + 2\{33,2\}\{23,3\} - \{33,1\}\frac{\partial}{\partial r}\log\sqrt{-g} - \{33,2\}\frac{\partial}{\partial \theta}\log\sqrt{-g} = 0$$

$$-\frac{\partial}{\partial r}\{44,1\} + 2\{44,1\}\{14,4\} - \{44,1\}\frac{\partial}{\partial r}\log\sqrt{-g} = 0.$$

Of the remaining equations, $G_{12} = 0$ gives

$$\{13,3\}\{23,3\} - \{12,2\}\frac{\partial}{\partial \theta}\log\sqrt{-g} = 0,$$

which disappears when the values of the symbols are substituted; and in the others there are no surviving terms.

Substituting from (28.5) and (28.32) the four equations give immediately

$$-\frac{1}{2}\lambda'' + \frac{1}{4}\lambda'^2 + \frac{1}{r^2} + \frac{1}{r^2} + \frac{1}{4}\nu'^2 + \left(\frac{1}{2}\lambda'' + \frac{1}{2}\nu'' - \frac{2}{r^2}\right) - \frac{1}{2}\lambda'\left(\frac{1}{2}\lambda' + \frac{1}{2}\nu' + \frac{2}{r}\right) = 0.$$

$$e^{-\lambda}(1 - r\lambda') - 2e^{-\lambda} + \cot^2\theta - \operatorname{cosec}^2\theta + re^{-\lambda}\left(\frac{1}{2}\lambda' + \frac{1}{2}\nu' + \frac{2}{r}\right) = 0,$$

$$\sin^2\theta\, e^{-\lambda}(1 - r\lambda') + (\cos^2\theta - \sin^2\theta) - 2\sin^2\theta\, e^{-\lambda} - 2\cos^2\theta + r\sin^2\theta\, e^{-\lambda}\left(\frac{1}{2}\lambda' + \frac{1}{2}\nu' + \frac{2}{r}\right) + \cos^2\theta = 0$$

$$-\frac{1}{2}e^{\nu-\lambda}(\nu'' + \nu'^2 - \nu'\lambda') + \frac{1}{2}e^{\nu-\lambda}\nu'^2 - \frac{1}{2}e^{\nu-\lambda}\nu'\left(\frac{1}{2}\lambda' + \frac{1}{2}\nu' + \frac{2}{r}\right) = 0.$$

These reduce to

$$\begin{aligned}
G_{11} &= \frac{1}{2}\nu'' - \frac{1}{4}\lambda\nu' + \frac{1}{4}\nu'^2 - \frac{\lambda'}{r} = 0, \\
G_{22} &= e^{-\lambda}\left(1 + \frac{1}{2}r(\nu' - \lambda')\right) - 1 = 0, \\
G_{33} &= \sin^2\theta\, e^{-\lambda}\left(1 + \frac{1}{2}r\,(\nu' - \lambda')\right) - \sin^2\theta = 0, \\
G_{44} &= e^{\nu-\lambda}\left(-\frac{1}{2}\nu'' + \frac{1}{4}\lambda'\nu' - \frac{1}{4}\nu'^2 - \frac{\nu'}{r}\right) = 0.
\end{aligned} \tag{28.6}$$

From the first and last equations $\lambda' = -\nu'$, and since both λ and ν must tend to zero at infinity $\lambda = -\nu$. The second and third equations (which are identical) then give

$$e^{\nu}(1 + r\nu') = 1.$$

Set $e^{\nu} = \gamma$, then

$$\gamma + r\gamma' = 1.$$

Whence

$$\gamma = 1 - \frac{2m}{r}, \tag{28.7}$$

where $2m$ is a constant of integration. m will later be identified with the mass of the particle in gravitational units. This solution satisfies the first and fourth equations, and, therefore, substituting in (28.22), we have as a possible expression for the line-element

$$ds^2 = -\gamma^{-1}\,dr^2 - r^2\,d\theta^2 - r^2\sin^2\theta\,d\varphi^2 + \gamma\,dt^2 \tag{28.8}$$

with $\gamma = 1 - 2m/r$.

It will be seen that the measured space around a particle is not Euclidean. Any actual measurement with our clock-scale gives the invariant quantity ds. If we lay our measuring-rod transversely, $ds = rd\theta$, so that our transverse measures are correct in this system of co-ordinates; but if we lay it radially, $ds = \gamma^{-\frac{1}{2}}\,dr$, and the measures need to be multiplied by $\gamma^{\frac{1}{2}}$ to give dr. Thus, referring our results to Euclidean space, we may say that a standard measuring rod contracts when turned from the transverse to the radial direction.

We could, of course, decide to treat the radial measures as correct, and apply corrections to the transverse measures. This amounts to substituting dr' for $\gamma^{-\frac{1}{2}}\,dr$ in (28.8), and using r' as the radial co-ordinate. It is impossible to say which form of (28.8) corresponds to our ordinary polar co-ordinates, since we have never hitherto had to pay attention to the ambiguity.

The possibility of using any function of r, instead of r, for the distance is connected with the fact that Einstein's equations amount to only 6 independent relations between the 10 g's. Consequently, quite apart from boundary conditions, there is a large amount of arbitrariness in choice of g's, i.e., of co-ordinates. The reader may meet elsewhere with different expressions for the line-element due to a particle. The one adopted here was first given by Schwarzschild.

For some purposes the following analogy is helpful. Instead of considering continuous space-time, consider that fundamentally we are dealing with an aggregate of points. With Galilean co-ordinates $x, y, z, t\sqrt{-1}$ the points are uniformly packed. Any measure that we make is really a counting of points, and a particle always moves so as to pass through the fewest possible points between any two positions on its path. Any mathematical transformation of these co-ordinates disturbs, without disordering, the distribution of the points in space; but it is meaningless so long as we consider only the points and not the arbitrary continuous space we place them in. In a gravitational field the points are disordered according to some definite law. We can evidently re-arrange them so that the number of points in the circumference of a circle is less than π times the number in the diameter (a circle being a geodesic on a hypersphere, which is a locus such that the minimum number of points between any point on it and a fixed point called the center is constant).

This representation, however, gives only imaginary time and therefore imaginary motions. When extended to real motions it becomes too complex to be of much help.

5 THE CRUCIAL PHENOMENA

29. The Equations of Motion of a Particle in the Gravitational Field

Denote the contravariant vector $\frac{\partial x_\sigma}{\partial s}$ by A^σ. Then by (22.5) its covariant derivative is

$$A^\sigma_\alpha = \frac{\partial}{\partial x_\alpha}\left(\frac{\partial x_\sigma}{\partial s}\right) + \{\alpha\beta, \sigma\}\frac{\partial x_\beta}{\partial s}.$$

Multiply this by $\partial x_\alpha/\partial s$, we have

$$A^\alpha A^\sigma_\alpha = \frac{\partial^2 x_\sigma}{\partial s^2} + \{\alpha\beta, \sigma\}\frac{\partial x_\alpha}{\partial s}\frac{\partial x_\beta}{\partial s},$$

showing that the right-hand side is a contravariant vector.

Consider the equations

$$\frac{\partial^2 x_\sigma}{\partial s^2} + \{\alpha\beta, \sigma\}\frac{\partial x_\alpha}{\partial s}\frac{\partial x_\beta}{\partial s} = 0, \quad (\sigma = 1, 2, 3, 4,), \tag{29}$$

since the left-side is a vector, the equations will be satisfied (or not) independently of the choice of co-ordinates. In Galilean co-ordinates, the second term vanishes, and the equations reduce to $\partial^2 x_\sigma/\partial s^2 = 0$, which are the equations of a straight line. Equation (29) is thus the general equation of the locus which in Galilean co-ordinates becomes a straight line.

The path of a particle in Galilean co-ordinates (i.e., under no forces) is a straight line. The equations (29) are accordingly the equations of motion of a particle referred to any axes, provided there is no permanent gravitational field. Further, since they contain only first derivatives of the g's, in accordance with §27, these equations of motion will hold also when there is a permanent gravitational field.

The equations must evidently correspond to the condition,

$$\int ds \quad \text{is stationary,}$$

and could have been deduced from it by the calculus of variations. The path of a particle is a geodesic in all cases. It should be noticed that $\int ds$ is not generally a *minimum*.

30. Using the values (28.5) of Christoffel's symbols, the equation of motion (29) for $\sigma = 2$ becomes

$$\frac{d^2\theta}{ds^2} - \cos\theta \sin\theta \left(\frac{d\varphi}{ds}\right)^2 + \frac{2}{r}\frac{dr}{ds}\frac{d\theta}{ds} = 0. \tag{30.12}$$

Choose co-ordinates so that the particle moves initially in the plane $\theta = \pi/2$; then $d\theta/ds = 0$ initially, and $\cos\theta = 0$, so that $d^2\theta/ds^2 = 0$. The particle therefore continues to move in this plane. The equations for $\sigma = 1, 3, 4$ are then

$$\frac{d^2r}{ds^2} + \frac{1}{2}\lambda'\left(\frac{dr}{ds}\right)^2 - r\,e^{-\lambda}\left(\frac{d\varphi}{ds}\right)^2 + \frac{1}{2}e^{\nu-\lambda}\,\nu'\left(\frac{dt}{ds}\right)^2 = 0 \tag{30.11}$$

$$\frac{d^2\varphi}{ds^2} + \frac{2}{r}\frac{dr}{ds}\frac{d\varphi}{ds} = 0 \tag{30.13}$$

$$\frac{d^2t}{ds^2} + \nu'\frac{dr}{ds}\frac{dt}{ds} = 0 \tag{30.14}$$

Integrating (30.13) and (30.14), we have

$$r^2\frac{d\varphi}{ds} = h, \tag{30.21}$$

$$\frac{dt}{ds} = ce^{-\nu} = \frac{c}{\gamma}, \tag{30.22}$$

where h and c are constants of integration.

Instead of troubling to integrate (30.11), we can use (28.8), which plays the part of an integral of energy, viz.,

$$\gamma^{-1}\left(\frac{dr}{ds}\right)^2 + r^2\left(\frac{d\varphi}{ds}\right)^2 - \gamma\left(\frac{dt}{ds}\right)^2 = -1. \tag{30.23}$$

From these three integrals,

$$\left(\frac{dr}{ds}\right)^2 + r^2\left(\frac{d\varphi}{ds}\right)^2 + (\gamma - 1)\frac{h^2}{r^2} - c^2 = -\gamma,$$

or substituting for γ its value (28.7)

$$\left(\frac{dr}{ds}\right)^2 + r^2\left(\frac{d\varphi}{ds}\right)^2 = (c^2 - 1) + \frac{2m}{r} + 2m\frac{h^2}{r^3}$$
$$\text{with} \quad r^2\frac{d\varphi}{ds} = h. \tag{30.3}$$

Compare these with the ordinary Newtonian equations for elliptic motion,

$$\left(\frac{dr}{dt}\right)^2 + r^2\left(\frac{d\varphi}{dt}\right)^2 = -\frac{m}{a} + \frac{2m}{r},$$
$$r^2\frac{d\varphi}{dt} = h. \tag{30.4}$$

To make them correspond we must take $c^2 = 1 - m/a$ where a is the major semiaxis of the orbit. The term $2mh^2/r^3$ represents a small additional effect not predicted by the Newtonian theory. Further, the quantity m, introduced as a constant of integration, is now identified as the mass of the attracting particle measured in gravitational units. With regard to the use of ds instead of dt in (30.3) it must be remembered that ds is the "proper time" for the moving particle, so it is permissible to take ds as corresponding to the time in making a comparison with Newtonian dynamics.

Mass, time and distance are all ambiguously defined in Newtonian dynamics, and in defining them for the present theory we have some freedom of choice, provided that our definition agrees with the Newtonian definition in the limiting case of a vanishing field of force.

31. *The Perihelion of Mercury*

The ratio m/a or m/r is very small in all practical applications. If we take 1 kilometre as the unit of length and time $\left(= \frac{1}{300{,}000}\text{ sec.}\right)$ then for the earth's orbit $a = 1.49 \times 10^8$ and the angular velocity $\omega = 6.64 \times 10^{-13}$. Hence the mass of the sun,

$$m = \omega^2\, a^3 = 1.47 \text{ kilometers.} \tag{31.1}$$

Thus for applications in the solar system m/r is of order 10^{-8} and it is easily seen that h^2/r^2 is of the same order. Also the difference between dt and ds is of order $10^{-8}\, ds$.

From (30.3) we have

$$\left(\frac{h}{r^2}\frac{dr}{d\varphi}\right)^2 + \frac{h^2}{r^2} = (c^2 - 1) + \frac{2m}{r} + \frac{2mh^2}{r^3},$$

or writing $u = 1/r$

$$\left(\frac{du}{d\varphi}\right)^2 + u^2 = \frac{c^2-1}{h^2} + \frac{2mu}{h^2} + 2mu^3.$$

Differentiating with respect to φ

$$\frac{d^2u}{d\varphi^2} + u = \frac{m}{h^2} + 3mu^2. \tag{31.2}$$

Since h^2u^2 is of order 10^{-8} we obtain an approximate solution by neglecting $3mu^2$. This is

$$u = \frac{m}{h^2}\Big(1 + e\,\cos(\varphi - \tilde{\omega})\Big), \tag{31.3}$$

as in Newtonian dynamics.

For a second approximation, we substitute this value of u in tie small term $3mu^2$, and (31.2) becomes

$$\frac{d^2u}{d\varphi^2} + u = \frac{m}{h^2} + \frac{3m^3}{h^4} + \frac{6m^3}{h^4}\, e\,\cos(\varphi - \tilde{\omega}) + \frac{3m^3\, e^2}{2h^4}\Big(1 + \cos 2(\varphi - \tilde{\omega})\Big). \tag{31.4}$$

Of the small additional terms the only one which can give appreciable effects is the term in $\cos(\varphi - \tilde{\omega})$, which is of the proper period to produce a continually increasing effect by resonance. It is well known that the particular integral of

$$\frac{d^2u}{d\varphi^2} + u = A\cos\varphi$$

is

$$u = \frac{1}{2}A\,\varphi\sin\varphi.$$

Hence this term gives a part of u,

$$u_1 = \frac{3m^3e}{h^4}\,\varphi\,\sin(\varphi - \tilde{\omega})$$

Adding this to (31.3) we have

$$\begin{aligned} u &= \frac{m}{h^2}\left(1 + e\cos(\varphi - \tilde{\omega}) + \frac{3m^2}{h^2}\varphi\, e\sin(\varphi - \tilde{\omega})\right) \\ &= \frac{m}{h^2}\left(1 + e\cos(\varphi - \tilde{\omega} - \delta\tilde{\omega})\right), \end{aligned}$$

where $\delta\tilde{\omega} = \frac{3m^2}{h^2}\varphi$, and $(\delta\tilde{\omega})^2$ is neglected.

Thus whilst the planet moves through one revolution, the perihelion advances a fraction of a revolution equal to

$$\frac{\delta\tilde{\omega}}{\varphi} = \frac{3m^2}{h^2} = \frac{3m}{\alpha(1-e^2)} = \frac{12\pi^2\alpha^2}{c^2T^2(1-e^2)},$$

where T is the period of the planet, and the velocity of light c has been reinstated.

For the four inner planets the numerical values of this predicted motion of the perihelion are (per century):

	$\delta\tilde{\omega}$	$e\delta\tilde{\omega}$
Mercury	+42".9	+8".82
Venus	8.6	0.05
Earth	3.8	0.07
Mars	1.35	0.13

The value of $e\delta\tilde{\omega}$ is given because this corresponds to the perturbation which can be measured. Clearly when e is vanishingly small it is not possible to detect observationally any change in the position of perihelion. The orbits of Venus and the Earth are nearly circular so that the predicted effect is too small to detect.

The following table gives the outstanding discrepancies between the present theory and observation for $e\delta\tilde{\omega}$ and δe (per century) with their probable errors. The secular changes δe are analogous to $e\delta\tilde{\omega}$; and the two perturbations may be regarded as the two rectangular components of a vector. In the last column we give the outstanding discrepancies of $e\delta\tilde{\omega}$ on the Newtonian theory; those of δe are, of course, unaltered.

	Einstein's Theory		Newtonian
	$e\delta\tilde{\omega}$	δe	$e\delta\tilde{\omega}$
Mercury	$-0''.58 \pm 0''.29$	$-0''.88 \pm 0''.33$	$+8''.24$
Venus	-0.11 ± 0.17	$+0.21 \pm 0.21$	-0.06
Earth	0.00 ± 0.09	$+0.02 \pm 0.07$	$+0.07$
Mars	$+0.51 \pm 0.23$	$+0.29 \pm 0.18$	$+0.64$

It will be seen that the famous large discordance of the perihelion of Mercury is removed by Einstein's theory. No other charge of importance is made except a slight improvement, for the perihelion of Mars. Of the eight residuals, four exceed the probable error, and none exceed three times the probable error, so that the agreement is very satisfactory.

It may be noticed that according to (31.4) the orbit is not exactly an ellipse, even apart from this progression of the apse. But this (unlike the motion of perihelion) has no observational significance, and merely arises from our particular choice of measurement of r. In any case the curve in non-Euclidean space, which is to be described as an ellipse, must be a matter o convention.

It will be found (putting $dr/ds = 0$ in (30.11)) that for a circular orbit Kepler's third law is exactly fulfilled. This again is not an observable fact. To compare it with observation we should have to consider the nature of the astronomical observations from which the direct value of the axis of the orbit is measured.

32. *Deflection of a Ray of Light.*

In the absence of a gravitational field the velocity of light is unity, so that

$$\left(\frac{dx}{dt}\right)^2 + \left(\frac{dy}{dt}\right)^2 + \left(\frac{dz}{dt}\right)^2 = 1.$$

Accordingly

$$ds^2 = -dx^2 - dy^2 - dz^2 + dt^2 = 0. \tag{32.1}$$

Hence for the motion of light $ds = 0$, and by the principle of equivalence this invariant equation must hold also in the gravitational field.

It may be of interest to notice that for an observer travelling with the light, $dx = dy = dz = 0$, so that $dt = ds = 0$. Hence, if man wishes to achieve immortality and eternal youth, all he has to do is to cruise about space with the velocity of light. He will return to the earth after what seems to him an instant to find many centuries passed away.

Setting $ds = 0$ in (28.8) we have (for motion in a plane)

$$\gamma^{-1}\left(\frac{dr}{dt}\right)^2 + \left(r\frac{d\varphi}{dt}\right)^2 = \gamma. \tag{32.2}$$

Hence if v is the velocity of light in a direction making an angle V with the radius vector,

$$v^2(\gamma^{-1}\cos^2 V + \sin^2 V) = \gamma,$$

whence

$$v = \gamma(1 - (1-\gamma)\sin^2 V)^{-1/2} \tag{32.3}$$

The velocity thus depends on the direction; but it must be remembered that this co-ordinate velocity is not the velocity found directly from measures at the point considered. When we determine the velocity by measures made in a small region, and use natural measure (i.e, $g_{\mu\nu}$, having the values (16.3) at that point), the measured velocity is necessarily unity.

Since it is inconvenient to have the velocity of light varying with direction, we shall slightly alter our co-ordinates. Set

$$r = r_1 + m. \tag{32.4}$$

Then, neglecting squares of m/r_1

$$r^2 = r_1^2\left(1 + \frac{2m}{r}\right) = \gamma^{-1}r_1^2.$$

Substituting in (32.2)

$$\left(\frac{dr_1}{dt}\right)^2 + \left(r_1\frac{d\varphi}{dt}\right)^2 = \gamma^2,$$

so that in these co-ordinates,

$$v = \gamma = 1 - \frac{2m}{r} \simeq 1 - \frac{2m}{r_1} \tag{32.5}$$

for all directions. We can now drop the suffix of r_1 .

By Huygens' principle the direction of the ray is determined by the condition that the time between two points is stationary for small variations of the path. The course of the ray will therefore depend only on the variation of velocity, and will be the same as in a Euclidean space filled with material of suitable refractive index. The necessary refractive index μ is given by

$$\mu = \frac{1}{v} = 1 + \frac{2m}{r}. \tag{32.6}$$

We thus see that the gravitational field round a particle will act like a converging lens. The path of a ray through a medium stratified in concentric spheres is given by,

$$\mu p = \text{const.} \tag{32.71}$$

where p is the perpendicular from the center on the tangent.

By (32.6) we have to this order of approximation,

$$\mu^2 = 1 + \frac{4m}{r}. \tag{32.72}$$

But (32.71) and (32.72) are the integrals of angular momentum and energy for the Newtonian motion of a particle with velocity μ under the attraction of a mass $2m$, the orbit being a hyperbola of semi-axis $2m$. This hyperbola, therefore, gives the path of the light. If the distance from the focus to the apse is R, we have

$$a = 2m,$$
$$a\,(e-1) = R,$$

so that

$$e = 1 + \frac{R}{2m} \doteq \frac{R}{2m}$$

and the very small angle between the asymptotes

$$= \frac{2}{\sqrt{(e^2-1)}} \doteq \frac{2}{e} \doteq \frac{4m}{R}.$$

Thus a ray of light travelling from $-\infty$ to $+\infty$ and passing at a distance R from a particle of mass m experiences a total deflection.

$$\alpha = \frac{4m}{R}. \tag{32.8}$$

For a star seen close to the limb of the sun, by (31.1) $m = 1.47$ km., and $R =$ sun's radius $= 697,000$ km. Hence

$$\alpha = 1''.74.$$

It is curious to notice the occurrence of the factor 2 (mass$= 2m$) in the dynamical analogy. The deflection is twice what we should obtain on the Newtonian theory for a particle moving through the gravitational field with the velocity of light. The path of a light ray is not a geodesic (or rather the notion of a geodesic fails for motion with the speed of light); it is determined by stationary values of $\int dt$ instead of $\int ds$.

It may also be noted that the velocity of light decreases as the light falls to the attracting body.

33. It is hoped to test this prediction by observations of stars near the limb of the sun during a total eclipse. If the answer should be in the affirmative, the question will arise whether this must be considered to confirm Einstein's law of gravitation, or whether the deflection is sufficiently accounted for by the simple hypothesis that the mass of the electromagnetic energy of light is subject to gravitation. The unexpected factor 2 suggests that the deflection on Einstein's theory will be double that which would result from the ordinary electromagnetic theory. It is worth while to examine this more closely.

Consider a tube of light of unit cross-section and length ds (Fig. 4). Let the inclination of the ray to the axis of x be φ Let g be the acceleration of the gravitational field directed along Oy. Let E be the energy per unit volume; and c be the velocity of light, which on the electromagnetic theory is absolutely constant.

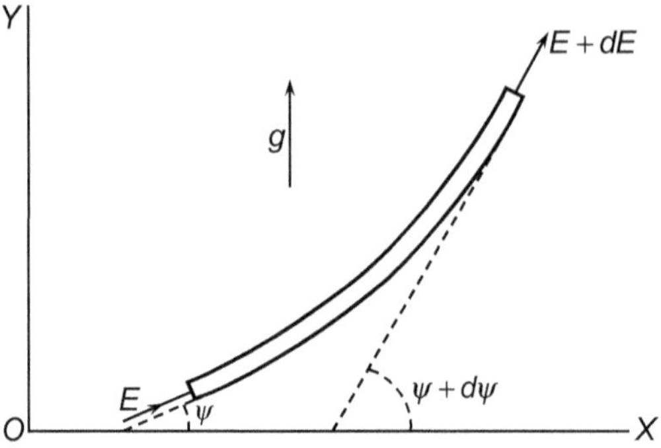

Fig. 4

Then the mass of electromagnetic energy E, according to electromagnetic theory (or by (7.85)), is E/c^2, so that if this is subject to gravity the momentum generated in the tube in unit time will be

$$\frac{E}{c^2} ds \cdot g \text{ along } Oy.$$

If the light is stopped by an absorbing screen placed perpendicular to the ray the radiation-pressure is numerically equal to E, showing that momentum E in the direction of the ray passes across a section of the tube in unit time. Thus, resolving in the x and y directions, the conservation of momentum gives

$$\begin{aligned} \frac{d}{ds}(E\cos\psi)\cdot ds &= 0, \\ \frac{d}{ds}(E\sin\psi)\cdot ds &= \frac{gE}{c^2}\,ds. \end{aligned} \tag{33.1}$$

Whence

$$\begin{aligned} \frac{dE}{ds}\cos\psi - E\sin\psi\frac{d\psi}{ds} &= 0, \\ \frac{dE}{ds}\sin\psi + E\cos\psi\frac{d\psi}{ds} &= \frac{gE}{c^2}. \end{aligned} \tag{33.2}$$

Eliminating dE/ds,

$$\frac{d\psi}{ds} = \frac{g}{c^2}\cos\psi. \tag{33.3}$$

The radius of curvature $ds/d\psi$ is thus $c^2/g\cos\psi$ which is exactly the same as for a material particle moving with velocity c in ordinary dynamics. This, as shown in the last paragraph, is only half the deflection indicated by Einstein's theory; and the experimental amount of the deflection should thus provide a crucial test.

34. *Displacement of Spectral Lines*

Consider an atom vibrating at any point of the gravitational field. It is a natural clock which ought to give an invariant measure of an interval δs; that is to say, the interval δs corresponding to one vibration of the atom is always the same, Let the atom be momentarily at rest in our system of co-ordinates (though subject to the acceleration of the field); then $dx = dy = dz = 0$, and by (15.3)

$$ds^2 = g_{44}\, dt^2$$

If then dt and dt' are the periods of two similar atoms vibrating at different parts of the field where the potentials are g_{44} and g'_{44}, respectively,

$$\sqrt{g_{44}} \cdot dt = \sqrt{g'_{44}} \cdot dt' \tag{34.1}$$

If t refers to an atom vibrating in the photosphere of the sun

$$g_{44} = 1 - \frac{2m}{R},$$

and if t' refers to an atom m a terrestrial laboratory, where g'_{44} is practically unity,

$$\frac{dt}{dt'} \simeq 1 + \frac{m}{R} = 1.00000212. \tag{34.2}$$

The solar atom thus vibrates more slowly, and its spectral lines will be displaced towards the red. The amount is equivalent to the Doppler displacement due to a velocity of 0.00000212, or in ordinary units 0.634 km. per sec. In the part of the spectrum usually investigated the displacement is about 0.008 tenth-metres.

The effect is of particular importance, because it has been claimed that the existence of this displacement is disproved by observations of the solar spectrum.[1] The difficulties of the test are so great that we may perhaps suspend judgment; but it would be idle to deny the seriousness of this apparent breakdown of Einstein's theory. We shall therefore consider the phenomenon from a more elementary point of view.

The phenomenon does not depend on the greater intensity of the field on the sun, but on the potential; and it can evidently occur in a uniform gravitational field. Consider an observer O in a uniform field of intensity g and two similar atoms A_1 and A_2, A_1 being close to the observer and A_2 at a distance a measured parallel to the field. The observer and his atoms will, of course, be falling with the acceleration g. Consider them all enclosed in a room which is also falling; then by the principle of equivalence O cannot detect any effect of the field, and he will therefore observe the same period of vibration T for both atoms. Now refer the phenomena to unaccelerated axes which coincide with the accelerated axes at the instant $t = 0$. The vibration emitted by A_2 at the time $t = 0$ will reach O at the time $t = a$ (the velocity of light being unity), by which time O will have acquired a velocity ga relative to the unaccelerated axes. He will, therefore, correct his observation of the period of A_2 for the Doppler effect of this velocity and deduce a true period $T/(1 - ga)$. The period of A_1 will require no correction, and will still be given, as T. Since ga is the difference of potential between A_1 and A_2 this agrees with (34.2).

[1] C. E. St. John, "Astrophysical Journal," Vol. 46, p. 249.

As an example of a varying field, consider an observer O at the origin of co-ordinates and an atom A at a distance r in a field of centrifugal force of potential $\Omega = \frac{1}{2}\omega^2 r^2$, the atom being at rest at the time of emission of the light, but subject to the acceleration of the field. Another way of stating the problem is that there is no field of force, and the atom is moving with velocity ωr at right angles to the radius vector at the time of emission of the light. But in that case the period of vibration is by §4 increased in the ratio

$$\begin{aligned}\beta &= (1 - \omega^2 r^2)^{-\frac{1}{2}} = (1 - 2\Omega)^{-\frac{1}{2}} \\ &= \frac{1}{\sqrt{g_{44}}} \text{ by } (16.2)\end{aligned}$$

as compared with the stationary atom. This again agrees with (34.1).

These verifications seem to leave little chance of evading the conclusion that a displacement of the Fraunhofer lines is a necessary and fundamental condition for the acceptance of Einstein's theory; and that if it is really non-existent, under conditions which strictly accord with those here postulated, we should have to reject the whole theory constructed on the principle of equivalence. Possibly a compromise might be effected by supposing that gravitation is an attribute only of matter in bulk and not of individual atoms; but this would involve a fundamental restatement of the whole theory.

If the displacement of the solar lines were confirmed, it would be the first *experimental* evidence that relativity holds for quantum phenomena.

6 The Gravitation of a Continuous Distribution of Matter

35. In the problems occurring in Nature our data give, not the distribution of the individual atoms, but the large-scale average distribution of density. This transition from discrete particles to the equivalent continuous medium occurs in the Newtonian theory of attractions, and involves the replacement of Laplace's equation $\nabla^2\varphi = 0$ by Poisson's equation $\nabla^2\varphi = -4\pi\rho$. We shall now find the corresponding modification of Einstein's equations $G_{\sigma\tau} = 0$.

The equations $G_{\sigma\tau} = 0$ are not linear in the g's, and consequently the fields of two or more particles are not strictly additive. But the deviations produced in the g's by any natural gravitational field are extremely small, so we shall neglect the product terms and treat the fields as superposable. It will be shown below that ultimately this approximation does not produce any inaccuracy in the application we have in view.

As in (32.4) we shall write $r = r_1 + m$ in (28.8) and neglect $(m/r)^2$. Then the line element in the field surrounding the particle is

$$ds^2 = -\gamma^{-1}\,(dr_1^2 + r_1^2\,d\theta^2 + r_1^2\sin^2\theta\,d\varphi^2) = \gamma\,dt^2. \tag{35.1}$$

We consider r_1 to be the actual radius vector, since the mode of measurement is arbitrary to this extent. Converting Into rectangular co-ordinates,

$$\begin{aligned} ds^2 &= -\gamma^{-1}(dx^2 + dy^2 + dz^2) + \gamma\,dt^2 \\ &= -\left(1 + \frac{2m}{r}\right)(dx^2 + dy^2 + dz^2) + \left(1 - \frac{2m}{r}\right)dt^2. \end{aligned} \tag{35.2}$$

The origin is now arbitrary, and r denotes the distance of the attracting particle from the element ds. The effects of a number of particles being additive to our order of approximation, we shall have for any number of particles at rest relative to the axes,

$$ds^2 = -(1 + 2\Omega)(dx^2 + dy^2 + dz^2) + (1 - 2\Omega)dt^2, \tag{35.3}$$

where $\Omega = \Sigma(m/r) =$ the Newtonian potential.

Consider a point O in the medium where the density is ρ, and with O as centre describe an infinitely small sphere. If we neglect the material inside the sphere, the equations of the gravitational field in free space will be satisfied at O, i.e., $G_{\sigma\tau} = 0$. Hence in calculating the values of $G_{\sigma\tau}$ at O we need only take account of the material inside the sphere. Accordingly in (35.3) Ω refers to the potential inside an infinitely small sphere of uniform density ρ.

Since $\frac{\partial\Omega}{\partial x}$, etc., vanish at O, we have only to take account of terms in (28.1) containing second derivatives of the g's; and the calculation of $G_{\sigma\tau}$ at O is quite simple. We have

$$\begin{aligned} G_{11} &= -\frac{\partial}{\partial x_1}\{11,1\} - \frac{\partial}{\partial x_2}\{11,2\} - \frac{\partial}{\partial x_3}\{11,3\} + \frac{\partial^2}{\partial x_1^2}\log\sqrt{-g} \\ &= -\frac{1}{2}g^{11}\frac{\partial^2 g_{11}}{\partial x_1^2} + \frac{1}{2}g^{22}\frac{\partial^2 g_{11}}{\partial x_2^2} + \partial 12 g_{33}\frac{\partial^2 g_{11}}{\partial x_3^2} + \frac{\partial^2}{\partial x_1^2}\log\sqrt{-g}, \end{aligned} \tag{35.4}$$

omitting 33 terms which vanish or cancel.

At O

$$g_{11} = g^{11} = -1, \quad g_{44} = g^{44} = 1 \tag{35.5}$$

and by (35.3)

$$\frac{\partial^2 g_{11}}{\partial x_1^2} = \frac{\partial^2 g_{44}}{\partial x_1^2} = -\frac{\partial^2}{\partial x_1^2}\log\sqrt{-g} = -2\frac{\partial^2\Omega}{\partial x_1^2}.$$

Hence substituting in (35.4)

$$\begin{aligned} G_{11} &= \frac{\partial^2\Omega}{\partial x_1^2} + \frac{\partial^2\Omega}{\partial x_2^2} + \frac{\partial^2\Omega}{\partial x_3^2} \\ &= -4\pi\rho \quad \text{by Poisson's equation.} \end{aligned}$$

Working out the other components similarly (with slight variations in the case of G_{44}) we find

$$G_{11} = G_{22} = G_{33} = G_{44} = -4\pi\rho. \tag{35.6}$$

The scalar

$$\begin{aligned} G &= g^{\sigma\tau}G_{\sigma\tau} \\ &= -G_{11} - G_{22} - G_{33} + G_{44} \\ &= 8\pi\rho. \end{aligned} \tag{35.7}$$

Now from the covariant tensor

$$-8\pi T_{\sigma\tau} = G_{\sigma\tau} - \frac{1}{2}g_{\sigma\tau}G. \tag{35.8}$$

We have by (35.6) and (35.7)

$$T_{44} = \rho,$$

and all other components vanish.

Having thus found the value of $T_{\sigma\tau}$ in this special system of coordinates we could find its general value by (19.31). It is, however, simpler to proceed as follows. If x_μ is a co-ordinate of a point in the material, consider the quantity,

$$\rho \frac{dx_\mu}{ds}\frac{dx_\nu}{ds}. \tag{35.91}$$

Since with respect to our special axes the material is at rest

$$\frac{dx_\mu}{ds} = 0 \quad (\mu = 1, 2, 3), \quad \text{and} \quad \frac{dx_\mu}{ds} = 1 \quad (\mu = 4).$$

Hence all the components of (35.91) vanish except for $\mu = \nu = 4$ for which the component is ρ – just like $T_{\sigma\tau}$. This, however, is a contravariant tensor[1] and (35.8) requires a covariant tensor.

We therefore form the associated covariant tensor (§20b)

$$T_{\sigma\tau} = \rho\, g_{\mu\sigma}\, g_{\nu\tau}\, \frac{dx_\mu}{ds}\frac{dx_\nu}{ds}, \tag{35.92}$$

which agrees with (35.91) in our special co-ordinates.

The equations (35.8) and (35.92) are in covariant form, and are true in one system, hence they are true in all possible systems of co-ordinates. They are the general equations of the gravitational field in a continuous medium.

An alternative form of (35.8) is readily obtained, viz.,

$$G_{\sigma\tau} = -8\pi\,(T_{\sigma\tau} - \frac{1}{2} g_{\sigma\tau} T), \tag{35-93}$$

where T is the associated scalar $g^{\sigma\tau}T_{\sigma\tau}$. (This follows since on inner multiplication of (35.8) by $g^{\sigma\tau}$ we obtain $G = 8\pi\, T$).

36. We thus find that in a continuous medium, $G_{\sigma\tau}$, instead of vanishing, is equal to a tensor expressing the content and state of motion of the medium at the point considered. On the equations here found we have two observations to make.

(1) A little consideration will show that notwithstanding the approximations made at various stages of the proof, the results are quite rigorous. It is clear that so far as the calculations for the infinitely small sphere surrounding O are concerned, we are justified in neglecting the product terms, since in the limit they will vanish compared with the linear terms. Another way of seeing this is to consider that $G_{\sigma\tau}$ involves only derivatives up to the second at the origin; and therefore we need only expand the g's in powers of r as far as r^2; but in our units ρ is of dimensions r^{-2}, and since the g's in rectangular co-ordinates are of zero dimensions, any terms involving ρ^2 would be of the form $\rho^2 r^4$, and therefore need not be retained. The effect of the gravitation of the matter outside the sphere is eliminated completely by our choice of coordinates. We choose them so that at O the g's have the values (16.3), i.e. we use "natural

[1] ρ is to be treated as an invariant. Whatever the axes chosen ρ is to be the density in natural measure as estimated by an observer moving with the matter.

measure". Since our axes move with the matter at O, the first derivatives of the g's (expressing the force) will not vanish unless the matter at O is moving with the acceleration of the field, which is not the case if there is any internal stress. These first derivatives are omitted from our equations after (35.3), because as already explained the external matter alone contributes nothing to $G_{\sigma\tau}$; further, the cross-terms are zero, because the first derivatives of the g's arising from the matter inside the sphere vanish. The result is thus rigorous, provided that in measuring the invariant density ρ we use natural measure, i.e., the mass and unit volume must be taken according to the direct measures made by an observer at O moving with the material there.

The argument may be summarized thus: $G_{\sigma\tau}$ consists of terms of types

$I_2 + E_2 + I_1^2 + I_1E_1 + E_1^2 +$ terms in $I_0 +$ terms in E_0, where I and E refer to the matter internal and external to the small sphere, and the suffixes refer to the order of the derivatives. Terms in I_1 vanish by the symmetry of the sphere; terms in I_0 vanish as the sphere is made infinitely small; terms in E_0 vanish because we use natural measure; the terms $E_2 + E_1^2$ vanish by Einstein's equations for free space. All that is left is I_2, and as the sphere is made infinitely small our determination of its value becomes rigorous.

(2) In replacing a molecular medium by a continuous medium, it is not sufficient to average the distribution of mass and mass-motion only; we must also represent somehow the internal motions. This is done by adding another property to the continuous medium – the pressure, or stress-system. The tensor $T_{\sigma\tau}$ will contain terms corresponding to the pressure; these are negligible in practical calculations of the gravitational field because the pressure is of order ρ times the Newtonian potential, i.e., of order ρ^2. The terms are, however, important in the general equations of momentum and energy, and we shall consider them more fully in the next paragraph.

37. In the dynamics of a continuous medium the most fundamental part is taken by the associated mixed tensor,

$$T_\mu^\nu = g^{\nu\alpha}T_{\mu\alpha} = g_{\mu\sigma}\Sigma\rho_0\frac{dx_\sigma}{ds}\frac{dx_\nu}{ds}, \tag{37.1}$$

where we have inserted the Σ in order to take account of the variety of internal motions, and have written ρ_0 for ρ in order to call attention to the fact that it represents the density in natural measure and not the density referred to the arbitrary axes chosen.

T_μ^ν may be called the energy-tensor, though it is actually an *omnmnium gatherum* of energy, mass, stress and momentum.

First consider the meaning of this tensor in the absence of a gravitational field, and accordingly choose Galilean axes. If u, v, w are the component velocities of the particles,

$$\frac{dx}{dt} = u, \quad \frac{dy}{dt} = v, \quad \frac{dz}{dt} = w, \quad \left(\frac{ds}{dt}\right)^2 = 1 - u^2 - v^2 - w^2 = \beta^{-2}. \tag{37.2}$$

But by (7.92) the density referred to the axes chosen is

$$\rho = \beta^2 \rho_0 = \frac{\rho_0}{\left(\frac{ds}{dt}\right)^2}.$$

Hence

$$T_\mu^\nu = g_{\mu\sigma}\, \Sigma\rho \frac{dx_\sigma}{dt}\frac{dx_\nu}{dt}. \tag{37.3}$$

Putting in the Galilean values of $g_{\mu\sigma}$, we have

$$T_\mu^\nu = \begin{array}{llll} -\Sigma\rho u^2, & -\Sigma\rho v u, & -\Sigma\rho w u, & \Sigma\rho u \\ -\Sigma\rho u v & -\Sigma\rho v^2, & -\Sigma\rho w v, & \Sigma\rho v \\ -\Sigma\rho u w, & -\Sigma\rho v w, & -\Sigma\rho w^2, & \Sigma\rho w \\ -\Sigma\rho u, & -\Sigma\rho v, & -\Sigma\rho w, & \Sigma\rho \end{array} \tag{37.4}$$

This tensor may be separated into two parts, the first referring to the motion u_0, v_0, w_0 , of the center of mass of the particles in an element, and the second to their internal motions, u_1, v_1, w_1 relative to the center of mass. With regard to the last part, $\Sigma\rho u_1 v_1$ represents the rate of transfer of $u-$momentum across unit area parallel to the $y-$plane, and is therefore equal to the stress usually denoted by p_x. Hence (37.4) becomes

$$T_\mu^\nu = \begin{array}{llll} -p_{xx} - \rho u_0^2, & -p_{yx} - \rho v_0 u_0, & -p_{zx} - \rho w_0 u_0, & \rho u_0 \\ -p_{xy} - \rho u_0 v_0, & -p_{yy} - \rho v_0^2, & -p_{zy} - \rho w_0 v_0, & \rho v_0 \\ -p_{xz} - \rho u_0 w_0, & -p_{yz} - \rho v_0 w_0, & -p_{zz} - \rho w_0^2, & \rho w_0 \\ -\rho u_0, & -\rho v_0, & -\rho w_0, & \rho \end{array} \tag{37.5}$$

where ρ is now the whole density referred to the axes chosen. Consider the equations

$$\frac{\partial}{\partial x_\nu} T_\mu^\nu = 0. \tag{37.6}$$

Taking $\mu = 4$, and using (37.5), we get the well-known equation of continuity

$$\frac{\partial(\rho u_0)}{\partial x} + \frac{\partial(\rho v_0)}{\partial y} + \frac{\partial(\rho w_0)}{\partial z} + \frac{\partial\rho}{\partial t} = 0. \tag{37.7}$$

Taking $\mu = 1$

$$\begin{aligned} -\left(\frac{\partial p_{xx}}{\partial x} + \frac{\partial p_{xy}}{\partial y} + \frac{\partial p_{xz}}{\partial z}\right) &= \frac{\partial(\rho u_0^2)}{\partial x} + \frac{\partial(\rho u_0 v_0)}{\partial y} + \frac{\partial(\rho u_0 w_0)}{\partial z} + \frac{\partial(\rho u_0)}{\partial t} \\ &= \rho\left(u_0\frac{\partial u_0}{\partial x} + v_0\frac{\partial u_0}{\partial y} + w_0\frac{\partial u_0}{\partial z} + \frac{\partial u_0}{\partial t}\right) \quad \text{using (37.7)} \\ &= \rho\frac{Du_0}{Dt}. \end{aligned} \tag{37.8}$$

Now (37.7) and (37.8) are the fundamental equations of hydrodynamics. By assuming Galilean axes we have neglected any extraneous body-forces, and so the term $-\rho X$, which occurs on the right side of (37.8) in the more general form of the equation, does not appear in this case.

The equation (37.6) is thus equivalent to the general equations of a fluid under no forces.

38. The equation $\partial T_\mu^\nu/\partial x_\nu = 0$ represents a law of conservation. Choose one of the co-ordinates x_4, as independent variable, and integrate the equation through a three-dimensional volume marked out in the other co-ordinates. This gives

$$\frac{\partial}{\partial x_4}\iiint T_\mu^4\, dx_1 dx_2 dx_3 = -\iiint\left(\frac{\partial T_\mu^1}{\partial x_1}+\frac{\partial T_\mu^2}{\partial x_2}+\frac{\partial T_\mu^3}{\partial x_3}\right) dx_1 dx_2 dx_3$$

$$= \text{the surface integral of the normal component of } (T_\mu^1, T_\mu^2, T_\mu^3).$$

If the volume is such as to include the whole of the material, T_μ^ν vanishes on the surface; the surface-integral therefore vanishes, and hence the volume integral of T_μ^4 remains constant. If the surface does not include all the matter, any change of its content of T_μ^4 occurs by a flux across the surface measured by $(T_\mu^1, T_\mu^2, T_\mu^3)$. It will be seen from (37.5) that for the axes there used T_μ^4 represents the negative momentum and the mass (or energy), and that T_μ^2, etc., represent the flux of these quantities. Equation (37.6) therefore gives the law of conservation of momentum and mass, as may be verified from the corresponding hydromechanical equations.

39. Equation (37.6) is the degenerate form for Galilean co-ordinates of the covariant equation

$$T_{\mu\nu}^\nu = 0 \tag{39.11}$$

where $T_{\mu\nu}^\nu$, is the (contracted) covariant derivative of T_μ^ν (see (22.7)). Equation (39.11) thus holds for Galilean co-ordinates, and it does not contain derivatives of the g's higher than the first. Hence by the principle of equivalence it holds generally, including the case of a permanent gravitational field.

Taking equation (35.8)

$$G_{\sigma\tau} - \frac{1}{2}g_{\sigma\tau}\, G = -8\pi\, T_{\sigma\tau},$$

multiply by $g^{\tau\nu}$. We obtain

$$G_\sigma^\nu - \frac{1}{2}g_\sigma^\nu\, G = -8\pi\, T_\sigma^\nu. \tag{39.12}$$

Take the covariant derivative of both sides, and contract it,

$$G_{\sigma\nu}^\nu - \frac{1}{2}g_\sigma^\nu \frac{\partial G}{\partial x_\nu} = -8\pi\, T_{\sigma\nu}^\nu = 0 \tag{39.13}$$

whence by (20.1)

$$G_{\sigma\nu}^\nu = \frac{1}{2}\frac{\partial G}{\partial x_\sigma}. \tag{39.14}$$

Clearly this equation will have to be an identity, and it may be verified analytically, using the values (26.3) of $G_{\mu\nu}$. For $\sigma = 1, 2, 3, 4$, this identity gives the four relations between Einstein's ten equations, which have already been mentioned as reducing the number of independent conditions to six.

Conversely, from the identity (39.14) we can deduce (39.11), and hence obtain the equations of hydromechanics and the law of conservation directly from Einstein's law of gravitation. Further, by applying the hydromechanical equations to an isolated particle, we obtain the equations of motion (29). The mass of a particle has been introduced first as a constant of integration, and afterwards identified with the gravitation-mass by determining the motion of a particle in its field; it now appears that it is also the inertia-mass, because it satisfies the law of conservation of mass and momentum, which gives the recognized definition of inertia.

It is startling to find that the whole of the dynamics of material systems is contained in the law of gravitation; at first sight gravitation seems scarcely relevant in much of our dynamics. But there is a natural explanation. A particle of matter is a singularity in the gravitational field, and its mass is the pole-strength of the singularity; consequently the laws of motion of the singularities must be contained in the field-equations, just as those of electromagnetic singularities (electrons) are contained in the electromagnetic field-equations. The fact that Einstein's law predicts these well-known properties of matter seems to be a valuable confirmation of this theory.

The general equation (39.11) enables us to pass from the equations of a fluid under no body forces to the equations of a fluid in a field of force. It can be simplified considerably. By (22.7)

$$T^{\nu}_{\mu\nu} = \frac{\partial T^{\nu}_{\mu}}{\partial x_{\nu}} - \{\nu\mu, \beta\}\, T^{\nu}_{\beta} + \{\beta\nu, \nu\}\, T^{\beta}_{\mu}. \tag{39.21}$$

By (26.25) the last term becomes

$$\frac{1}{\sqrt{-g}}\,\frac{\partial}{\partial x_{\beta}}\left(\sqrt{-g}\right) T^{\beta}_{\mu}. \tag{39.22}$$

The second term is equal to

$$\begin{aligned} &- [\nu\mu, \varepsilon]\, g^{\varepsilon\beta}\, g^{\nu\alpha}\, T_{\alpha\beta} \\ &= -\frac{1}{2}\,\frac{\partial g_{\nu\epsilon}}{\partial x_{\mu}}\, g^{\alpha\beta}\, g^{\nu\alpha}\, T_{\alpha\beta}, \end{aligned}$$

since the other two terms cancel on summation,

$$= \frac{1}{2}\,\frac{\partial g^{\alpha\beta}}{\partial x_{\mu}}\, T_{\alpha\beta}. \tag{39.3}$$

This last result follows, since

$$g^{\nu\alpha}\, g_{\nu\epsilon} = 0 \text{ or } 1,$$

so that

$$g^{\nu\alpha}\, dg_{\nu\varepsilon} + g_{\nu\varepsilon}\, dg^{\nu\alpha} = 0.$$

Multiply by $g^{\varepsilon\beta}$ and use (20.15), we obtain

$$g^{\varepsilon\beta}\, g^{\nu\alpha}\, dg_{\nu\varepsilon} = -dg^{\alpha\beta}. \tag{39.4}$$

Hence inserting (39.22) and (39.3) in (39.21), we have

$$\frac{1}{\sqrt{-g}}\frac{\partial}{\partial x_\nu}\left(\sqrt{-g}\, T^\nu_\mu\right) = -\frac{1}{2}\frac{\partial g^{\alpha\beta}}{\partial x_\mu} T_{\alpha\beta}. \tag{39.45}$$

This equation has its simplest interpretation when we choose co-ordinates, so that $\sqrt{-g} = 1$, that is to say, the volume of a four-dimensional element is to be the same in co-ordinate measure as in natural measure. Owing to the considerable freedom of choice of co-ordinates, allowed by Einstein's equations, it is always possible to do this. In that case (39.45) becomes

$$\frac{\partial}{\partial x_\nu}\left(T^\nu_\mu\right) = -\frac{1}{2}\frac{\partial g^{\alpha\beta}}{\partial x_\mu} T_{\alpha\beta}. \tag{39.5}$$

Comparing this with (37.6), which holds when there is no field of force, we see that the term on the right represents the momentum and energy transferred from the gravitational field to the material system. As a first approximation (retaining only $T_{44} = \rho$ and $g_{44} = 1 - 2\Omega$) we see that it gives, for $\mu = 1, 2, 3$, the terms $\rho X, \rho Y, \rho Z$ of the usual hydrodynamical equations, which were omitted in (37.8).

40. Propagation of Gravitation

The velocity of light being a fundamental relation between the measures of time and space, we may expect the strains representing a varying gravitational field to be propagated with this velocity. We shall show how to derive the equations exhibiting the propagation.

In the theory of sound, the general equation of disturbances propagated with unit velocity is

$$\Box\varphi \equiv \left(\frac{\partial^2}{\partial t^2} - \frac{\partial^2}{\partial x^2} - \frac{\partial^2}{\partial y^2} - \frac{\partial^2}{\partial z^2}\right)\varphi = \Phi, \tag{40.11}$$

where Φ is zero except at the source of the disturbance. The general solution is

$$\varphi = \frac{1}{4\pi}\iiint \frac{\Phi'}{r'}\, dV', \tag{40.12}$$

the integral being taken through the volume occupied by the source of disturbance, and the value of Φ' taken for a time $t - r'$ where r' is the distance of the volume dV' from the point considered. Thus φ is a retarded potential, and (40.12) exhibits the effect as delayed by propagation.

In the case of sound the velocity depends to a slight extent on the amplitude, and (40.11) is only strictly true if the square of φ is negligible. Similarly the velocity of light depends to a slight extent on the gravitational field (§32); consequently we can only expect to obtain an equation of this form if we neglect the square of the disturbance, so that the equation become linear.

The origin of gravitational waves must be attributed to moving matter; and, since $G_{\mu\nu}$ vanishes except in a region occupied by matter, we may take $G_{\mu\nu}$ as the analogue of Φ. We shall examine whether the disturbance can be represented by a quantity $h_{\mu\nu}$ satisfying

$$\Box h_{\mu\nu} = 2G_{\mu\nu}, \tag{40.21}$$

where the exact significance of $h_{\mu\nu}$ is yet to be found. We shall regard $h_{\mu\nu}$ as a small quantity of the first order; the deviations of the $g_{\mu\nu}$ from their Galilean values will also be of the first order. Small quantities of the second order will be neglected. If, as usual

$$h_\mu^\sigma = g^{\nu\sigma}\, h_{\mu\nu},$$

and

$$h = g^{\mu\nu}\, h_{\mu\nu}.$$

Then, multiplying (40.21) successively by $g^{\nu\sigma}$ and $g^{\mu\nu}$, we have to this approximation[2]

$$\Box h_\mu^\sigma = 2\, G_\mu^\sigma \tag{40.22}$$

and

$$\Box h = 2\, G \tag{40.23}$$

Hence

$$\begin{aligned}\Box\big(h_\mu^\sigma - \frac{1}{2}\, g_\mu^\sigma\, h\big) &= 2\big(G_\mu^\sigma - \frac{1}{2}\, g_\mu^\sigma\, G\big)\\ &= -16\pi\, T_\mu^\sigma \quad \text{by (39.12)}.\end{aligned}$$

To the present approximation (37.6) holds, so that

$$\Box\left(\frac{\partial}{\partial x_\sigma}\, h_\mu^\sigma - \frac{1}{2}\, g_\mu^\sigma\, \frac{\partial h}{\partial x_\sigma}\right) = -16\pi\, \frac{\partial}{\partial x_\sigma}\, T_\mu^\sigma = 0.$$

Having regard to boundary conditions, the solution is clearly

$$\begin{aligned}\frac{\partial}{\partial x_\sigma}\, h_\mu^\sigma &= \frac{1}{2}\, g_\mu^\sigma\, \frac{\partial h}{\partial x_\sigma}\\ &= \frac{1}{2}\, \frac{\partial h}{\partial x_\mu}.\end{aligned} \tag{40.3}$$

Consider the expression

$$-\frac{1}{2}\, \frac{\partial}{\partial x_\alpha}\left\{g^{\alpha\beta}\left(\frac{\partial h_{\nu\beta}}{\partial x_\mu} + \frac{\partial h_{\mu\beta}}{\partial x_\nu} - \frac{\partial h_{\mu\nu}}{\partial x_\beta}\right)\right\} + \frac{1}{2}\, \frac{\partial}{\partial x_\mu}\left(g^{\mu\nu}\, \frac{\partial h_{\alpha\nu}}{\partial x_\nu}\right), \tag{40.4}$$

[2]The $g^{\mu\nu}$ behave as constants until we reach equation (40.5), because their derivatives, which are small quantities of the first order, only appear in combination with the small quantities $h_{\mu\nu}$ or $G_{\mu\nu}$. The $g^{\mu\nu}$ accordingly pass freely under the differential operators.

which to our approximation

$$= -\frac{1}{2}\frac{\partial^2 h_\nu^\alpha}{\partial x_\alpha \partial x_\mu} - \frac{1}{2}\frac{\partial^2 h_\mu^\alpha}{\partial x_\alpha \partial x_\nu} + \frac{1}{2} g^{\alpha\beta}\frac{\partial^2 h_{\mu\nu}}{\partial x_\alpha \partial x_\beta} + \frac{1}{2}\frac{\partial^2 h}{\partial x_\mu \partial x_\nu}.$$

By (40.3) the first two terms cancel with the last, and for Galilean values of $g^{\alpha\beta}$ the third term is simply

$$\frac{1}{2}\Box h_{\mu\nu}.$$

Thus by (40.21) the expression (40.4) reduces to $G_{\mu\nu}$.

Neglecting squares of small quantities, $G_{\mu\nu}$ (26.3) reduces to

$$-\frac{\partial}{\partial x_\alpha}\{\mu\nu, \alpha\} + \frac{\partial^2}{\partial x_\mu \partial x_\nu}\log\sqrt{-g}$$
$$= -\frac{1}{2}\frac{\partial}{\partial x_\alpha}\left\{g^{\alpha\beta}\left(\frac{\partial g_{\nu\beta}}{\partial x_\mu} + \frac{\partial g_{\mu\beta}}{\partial x_\nu} - \frac{\partial g_{\mu\nu}}{\partial x_\beta}\right)\right\} + \frac{1}{2}\frac{\partial}{\partial x_\mu}\left(g^{\mu\nu}\frac{\partial g_{\mu\nu}}{\partial x_\nu}\right). \quad (40.5)$$

Comparing (40.4) and (40.5) we see that the h's must be equal to the g's – or rather since the h's have been treated as small quantities, they must be the deviations of the g's from their constant Galilean values. Writing $\delta_{\mu\nu}$ for the Galilean values of $g_{\mu\nu}$ (16.3), then

$$g_{\mu\nu} = \delta_{\mu\nu} + h_{\mu\nu}, \quad (40.6)$$

and $h_{\mu\nu}$ satisfies the equation of wave-propagation (40.21).

By (40.12) the solution of the propagation equation is

$$\begin{aligned} h_{\mu\nu} &= \frac{1}{2\pi}\iiint \frac{G'_{\mu\nu}}{r'}\,dV' \\ &= -4\iiint \frac{T'_{\mu\nu}}{r'}\,dV' + 2\delta_{\mu\nu}\iiint \frac{T'}{r'}\,dV'. \end{aligned} \quad (40.7)$$

This can be used for the practical calculation of $g_{\mu\nu}$ due to an arbitrary distribution of moving matter. It is necessary, as in the corresponding calculation of retarded electromagnetic potentials, to allow for the variation of $t - r'$ from point to point of the body; the boundary of dV' does not coincide with the limits of the body at any one instant. Thus for a particle of mass m, we have[3]

$$\iiint \frac{T'_{44}}{r'}\,dV' = \iiint \frac{\rho'}{r'}\,dV' = \frac{m}{[r\,(1 - v_r)]},$$

where v_r is the velocity in the direction of r, and the square bracket indicates retarded values. As is well known $[r(1 - v_r)]$ is to the first order equal to the unretarded distance r, so that notwithstanding the finite velocity of propagation the force is directed approximately towards the contemporaneous position of the

[3]See, for example, Lorentz, "The Theory of Electrons," p. 254; or Cunningham, "The Principle of Belativity," p 108.

attracting body. It was lack of knowledge of this compensation which led Laplace and many following him to state that the velocity of gravitation must far exceed the velocity of light.

The practical application of these formulae is, however, very limited. In a natural system (e.g., the solar system) the relative velocities (u) are due to the gravitational field and u^2 is a small quantity of the first order. Consequently our approximation is not good enough to take account of T_{11}, T_{12} etc., in natural systems; it can only include components with suffix 4.[4] The fact is that the whole idea of propagation from a point-source is an abstraction; actually the motion of the source, or singularity, is but the symbol of the changes occurring in all parts of the field; we cannot say whether the motion is the cause or effect of the gravitational waves.

The present solution is a particular solution. It gives unique values of the g_{μ}, but these may, of course, be subjected to arbitrary transformations.

[4]For the higher approximations needed in the problems of the solar system, see De Sitter, "Monthly Notices," Dec. 1916.

7 The Principle of Least Action

41. *Lagrange's Equations*

We shall again restrict the choice of co-ordinates so that $\sqrt{-g} = 1$. Einstein's equations (26.3) for the field in free space then becomes simplified to

$$G_{\mu\nu} \equiv -\frac{\partial}{\partial x_\alpha}\{\mu\nu,\alpha\} + \{\mu\beta,\alpha\}\{\nu\alpha,\beta\} = 0. \tag{41.1}$$

We shall regard $g^{\mu\nu}$ as a generalized co-ordinate (q), and $x_1, x_2, x_3 x_4$ as independent variables – a four-dimensional time. Writing $g^{\mu\nu}_\alpha$ for $\partial g^{\mu\nu}/\partial x_\alpha$, which will then be a generalized velocity ($\dot{q}$), we shall show that equations (41.1) can be expressed in the Lagrangian form.

$$G_{\mu\nu} = \frac{\partial}{\partial x_\alpha}\left(\frac{\partial L}{\partial g^{\mu\nu}_\alpha}\right) - \frac{\partial L}{\partial g^{\mu\nu}} = 0, \tag{41.2}$$

where

$$L = g^{\mu\nu}\{\mu\beta,\alpha\}\{\nu\alpha,\beta\} \tag{41.3}$$

it being understood that the $g_{\mu\nu}$ are expressed as functions of the $g^{\mu\nu}$.

We have from (41.3)

$$\delta L = \{\mu\beta,\alpha\}\{\nu\alpha,\beta\}\,\delta g^{\mu\nu} + 2g^{\mu\nu}\{\mu\beta,\alpha\}\,\delta\{\nu\alpha,\beta\},$$

since in the last term μ and ν are dummies.

$$= -\{\mu\beta,\alpha\}\{\nu\alpha,\beta\}\,\delta g^{\mu\nu} + 2\{\mu\beta,\alpha\}\,\delta[g^{\mu\nu}\{\nu\alpha,\beta\}].$$

But

$$\delta\left[g^{\mu\nu}\{\nu\alpha,\beta\}\right] = \frac{1}{2}\,\delta\left[g^{\mu\nu}\,g^{\beta\lambda}\left(\frac{\partial g_{\nu\lambda}}{\partial x_\alpha} + \frac{\partial g_{\alpha\lambda}}{\partial x_\nu} - \frac{\partial g_{\alpha\nu}}{\partial x_\lambda}\right)\right].$$

The last two terms in the bracket will cancel in the summation after inner multiplication by $\{\mu\beta,\alpha\}$, because μ and β, ν and λ are interchangeable simultaneously. Also by (39.4)

$$g^{\mu\nu}\,g^{\beta\lambda}\,\frac{\partial g_{\nu\lambda}}{\partial x_\alpha} = -\frac{\partial g^{\mu\beta}}{\partial x_\alpha}.$$

Hence

$$\delta L = -\{\mu\beta, \alpha\}\,\{\nu\alpha, \beta\}\,\delta g^{\mu\nu} - \{\mu\beta, \alpha\}\,\delta g^{\mu\beta}_{\alpha}.$$

Therefore

$$\begin{aligned} \frac{\partial L}{\partial g^{\mu\nu}_{\alpha}} &= -\{\mu\nu, \alpha\}, \\ \frac{\partial L}{\partial g^{\mu\nu}} &= -\{\mu\beta, \alpha\}\,\{\nu\alpha, \beta\} \end{aligned} \tag{41.4}$$

showing that (41.1) and (41.2) are equivalent.

As in ordinary dynamics, Lagrange's equations are equivalent to

$$\int L\,d\tau \quad \text{is stationary} \tag{41.5}$$

for variations of $g^{\mu\nu}$, $d\tau$ being the four-dimensional element of volume, here representing the independent variable. It must be remembered that the variations are limited by the constraint $\sqrt{-g} = 1$.

42. *Principle of Least Action*[1]

Following out the dynamical analogy $\partial L/\partial g^{\mu\nu}_{\alpha}$ or $\partial L/\partial \dot{q}$ is to be regarded as a momentum (p). The system is dynamically of the simplest kind, since L does not contain the "time" x_μ, explicitly, and it is a homogeneous quadratic function of the "velocities". By the properties of homogeneous functions

$$2L = \Sigma \dot{q}\,\frac{\partial L}{\partial \dot{q}} = \Sigma \dot{q} p.$$

Since $(p\dot{q} + q\dot{p})$ is a perfect differential,

$$\int \Sigma(\dot{p}q + q\dot{p})\,d\tau$$

will be equal to a surface integral; and it will, therefore be stationary for variations of $g^{\mu\nu}$ (the variations as usual being supposed to vanish at the boundary).

Thus

$$\delta \int \Sigma q\dot{p}\,d\tau = -\delta \int \Sigma \dot{q} p\,d\tau = -2\delta \int L\,d\tau. \tag{42.1}$$

Hence, if we write

$$H = L + \Sigma q\dot{p} \tag{42.2}$$

by (41.5) and (42.1)

$$\int H\,d\tau \quad \text{is stationary.} \tag{42.3}$$

By (41.4)

$$\Sigma q\dot{p} = -g^{\mu\nu}\,\frac{\partial}{\partial x_\alpha}\,\{\mu\nu, \alpha\}.$$

[1]The strict analogue of the principle of least action is the stationary property of $\int \Sigma qp\,d\tau$. The restriction in dynamics that the energy is not to be varied corresponds to $\sqrt{-g} = 1$. (Cf. §43).

Hence (42.2), (41.3) and (41.1) give

$$H = g^{\mu\nu}\, G_{\mu\nu} = G.$$

We can therefore write the result (42.3) thus

$$\int G\,\sqrt{-g}\,d\tau \quad \text{is stationary} \tag{42.4}$$

since $\sqrt{-g} = 1$.

But G and $\sqrt{-g}\,d\tau$ are invariants (20.3); so that (42.4) has no reference to any particular choice of co-ordinates, and the restriction $\sqrt{-g} = 1$ can now be removed. It is thus a more general result than (41.5).

43. *Energy of the Gravitational Field*

Reverting to the restriction $\sqrt{-g} = 1$, multiply (41.2) by $g^{\mu\nu}_{\beta}$

$$g^{\mu\nu}_{\beta}\, G_{\mu\nu} = g^{\mu\nu}_{\beta}\, \frac{\partial}{\partial x_\alpha}\left(\frac{\partial L}{\partial g^{\mu\nu}_{\alpha}}\right) - \frac{\partial L}{\partial g^{\mu\nu}}\,\frac{\partial g^{\mu\nu}}{\partial x_\beta}. \tag{43.1}$$

But

$$\frac{\partial L}{\partial x_\beta} = \frac{\partial L}{\partial g^{\mu\nu}}\,\frac{\partial g^{\mu\nu}}{\partial x_\beta} + \frac{\partial L}{\partial g^{\mu\nu}_{\alpha}}\,\frac{\partial g^{\mu\nu}_{\alpha}}{\partial x_\beta}. \tag{43.2}$$

Remembering that

$$\frac{\partial}{\partial x_\beta}\, g^{\mu\nu}_{\alpha} = \frac{\partial^2 g^{\mu\nu}}{\partial x_\beta \partial x_\alpha} = \frac{\partial}{\partial x_\alpha}\, g^{\mu\nu}_{\beta},$$

we have, adding (43.1) and (43.2),

$$g^{\mu\nu}_{\beta}\, G_{\mu\nu} = \frac{\partial}{\partial x_\alpha}\left(g^{\mu\nu}_{\beta}\,\frac{\partial L}{\partial g^{\mu\nu}_{\alpha}}\right) - \frac{\partial L}{\partial x_\beta} \tag{43.3}$$

$$= -16\pi\,\frac{\partial}{\partial x_\alpha}\, t^{\alpha}_{\beta}, \tag{43.4}$$

where

$$-\,16\pi\, t^{\alpha}_{\beta} = g^{\mu\nu}_{\beta}\,\frac{\partial L}{\partial g^{\mu\nu}_{\alpha}} - g^{\alpha}_{\beta}\, L. \tag{43.5}$$

We have used the property of g^{α}_{β} as substitution operator. The quantity t^{α}_{β} defined by (43.5) is the analogue of the Hamiltonian integral of energy, $\Sigma \dot{q}\,\frac{\partial L}{\partial \dot{q}} - L$. In free space $G_{\mu\nu} = 0$ and (43.4) becomes

$$\frac{\partial}{\partial x_\alpha}\, t^{\alpha}_{\beta} = 0 \tag{43.6}$$

showing that t^{α}_{β} is conserved (§38).

When matter is present (43.4) gives

$$-16\pi\,\frac{\partial t^{\alpha}_{\beta}}{\partial x_\alpha} = \frac{\partial g^{\mu\nu}}{\partial x_\beta}\, G_{\mu\nu}$$

$$= \frac{\partial g^{\mu\nu}}{\partial x_\beta}\left(G_{\mu\nu} - \frac{1}{2}\, g_{\mu\nu}\, G\right),$$

since, when $g = -1$, $g_{\mu\nu}\, dg^{\mu\nu} = 0$.

Hence by (35.8)

$$\begin{aligned}\frac{\partial}{\partial x_\alpha}\, t^\alpha_\beta &= \frac{1}{2}\,\frac{\partial g^{\mu\nu}}{\partial x_\beta}\, T_{\mu\nu} \\ &= -\frac{\partial}{\partial x_\alpha}\, T^\alpha_\beta \quad \text{by (39.5).}\end{aligned} \tag{43.7}$$

Therefore

$$\frac{\partial}{\partial x_\alpha}\,(T^\alpha_\beta + t^\alpha_\beta) = 0. \tag{43.8}$$

This is the law of conservation in the general case when there is interaction between matter and the gravitational field. We see that the changes of energy and momentum of the matter can be regarded as due to a transfer from or to the gravitational field, the total amount being conserved. We have, in fact, traced the disappearing portion of the material tensor T^ν_μ and shown that it reappears as the quantity t^ν_μ belonging to the gravitational field.

In order to represent the phenomena in this way we have had to restrict the choice of co-ordinates by keeping the volume of a region of space-time invariant ($\sqrt{-g} = 1$). Otherwise the equation takes the more general form (39.11) which cannot immediately be interpreted as a law of conservation. It should be noted that, unlike T^α_β, the quantity t^α_β is not strictly a tensor.

44. *The Method of Hilbert and Lorentz*

An alternative method of deriving the fundamental equations of this theory is based on the postulate that all the laws of mechanics can be summed up in a generalized principle of stationary action, viz.,

$$\delta \int (H_1 + H_2 + H_3 + \ldots)\, \sqrt{-g}\, d\tau = 0. \tag{44.1}$$

Here H_1, H_2, H_3 are invariants[2] involving, respectively, the parameters describing the gravitational field, the electromagnetic field, and the material system. If we consider matter and radiation in bulk we may add a fourth, term involving the entropy, so as to bring in thermodynamical phenomena, and so on. The variations are taken with respect to these parameters, their values at the boundary of integration being kept constant.

It is well known that the laws of mechanics of matter and of electrodynamics can be expressed in this form, so that we are here chiefly concerned with H_1. We already know from (42.4) that Einstein's theory is given by $H_1 = G$. Now G is, in fact, the principle invariant of the quadratic form $g_{\mu\nu}\, dx_\mu dx_\nu$ viz., the Gaussian invariant of curvature. This aspect of the theory seems to eliminate any element of arbitrariness which may have been felt when we fixed on the contracted Riemann-Christoffel tensor for the law of gravitation.

[2]Invariant because the equation must hold in all systems of coordinates, and we already know that the factor $\sqrt{-g}\, d\tau$ is invariant.

To interpret G as a curvature, consider a surface drawn in space of five dimensions, whose equation referred to the lines of curvature and the normal (z) at a point on it may be written

$$2z = k_1x_1^2 + k_2x_2^2 + k_3x_3^2 + k_4x_4^2 + \text{higher powers.} \tag{44.2}$$

wherek_1, k_2, k_3, k_4 are the reciprocals of the principle radii of curvature.

Then

$$ds^2 = dz^2 + \Sigma dx_1^2.$$

Eliminating z by (44.2)

$$ds^2 = (1 + k_1^2x_1^2)\, dx_1^2 + \ldots + 2k_1k_2x_1x_2dx_1dx_2 + \ldots \tag{44.3}$$

Hence at the origin,

$$g_{\mu\mu} = 1, \quad g_{\mu\nu} = 0\ (\mu \neq \nu), \quad \frac{\partial g_{\mu\nu}}{\partial x_\sigma} = 0.$$

The only surviving terms in $G = g^{\mu\nu}\, G_{\mu\nu}$ are

$$-g^{\mu\mu}\, \frac{\partial}{\partial x_\rho}\, \{\mu\mu, \rho\} + g^{\mu\mu}\, \frac{\partial^2}{\partial x_\mu^2}\, (\log \sqrt{-g}).$$

We easily find that

$$G = -2(k_1k_2 + k_2k_3 + k_3k_1 + k_1k_4 + k_2k_4 + k_3k_4). \tag{44.4}$$

In three dimensions we have only two curvatures, and k_1, k_2 is known as Gauss's measure of curvature, i.e., the ratio of the solid angle contained by the normals round the perimeter of an element to the area of the element. The expression (44.4) is a generalization of this invariant to five dimensions.

The curvature G in ordinary matter is quite considerable. In water the curvature is the same as that of a spherical space of radius 570,000,000 km. Presumably, if a globe of water of this radius existed, there would not be room in space for anything else.

45. *Electromagnetic Equations*

The electromagnetic field is described by a covariant vector $\varkappa_\mu$. In Galilean co-ordinates,

$$\varkappa_\mu = (-F, -G, -H, \Phi), \tag{45.1}$$

where F, G, H is the vector potential and Φ the scalar potential of the ordinary theory.

If $\varkappa_{\mu\nu}$ is the covariant derivative of $\varkappa_\mu$, we have by (22.2)

$$\begin{aligned}\frac{\partial \varkappa_\mu}{\partial x_\nu} - \frac{\partial \varkappa_\nu}{\partial x_\mu} &= \varkappa_{\mu\nu} - \varkappa_{\nu\mu} = \text{a covariant tensor,}\\ &= F_{\mu\nu}, \text{ say.}\end{aligned} \tag{45.2}$$

The electric and magnetic forces are given in the electromagnetic theory by

$$X = -\frac{\partial \Phi}{\partial x} - \frac{\partial F}{\partial t}, \qquad \alpha = \frac{\partial H}{\partial y} - \frac{\partial G}{\partial z}, \tag{45.3}$$

i.e.,

$$X = \frac{\partial \varkappa_1}{\partial x_4} - \frac{\partial \varkappa_4}{\partial x_1}, \qquad \alpha = \frac{\partial \varkappa_2}{\partial x_3} - \frac{\partial \varkappa_3}{\partial x_2}.$$

Hence by (45.2) the value of $F_{\mu\nu}$ in Galilean co-ordinates is

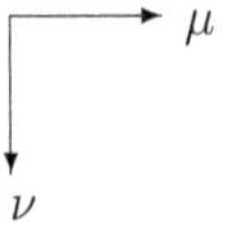

$$F_{\mu\nu} = \begin{matrix} 0 & -\gamma & \beta & -X \\ \gamma & 0 & -\alpha & -Y \\ -\beta & \alpha & 0 & -Z \\ X & Y & Z & 0 \end{matrix} \tag{45.41}$$

and the associated contravariant tensor, $F^{\mu\nu} = g^{\mu\alpha}\, g^{\nu\beta}\, F_{\alpha\beta}$ is,

$$F^{\mu\nu} = \begin{matrix} 0 & -\gamma & \beta & X \\ \gamma & 0 & -\alpha & Y \\ -\beta & \alpha & 0 & Z \\ -X & -Y & -Z & 0 \end{matrix} \tag{45.42}$$

We can now express Maxwell's equations in covariant form. In the ordinary theory they are

$$\frac{\partial Z}{\partial y} - \frac{\partial Y}{\partial z} = -\frac{\partial \alpha}{\partial t}, \quad \frac{\partial X}{\partial z} - \frac{\partial Z}{\partial x} = -\frac{\partial \beta}{\partial t}, \quad \frac{\partial Y}{\partial x} - \frac{\partial X}{\partial y} = -\frac{\partial \gamma}{\partial t}, \tag{45.51}$$

$$\frac{\partial \gamma}{\partial y} - \frac{\partial \beta}{\partial z} = \frac{\partial X}{\partial t} + u, \quad \frac{\partial \alpha}{\partial z} - \frac{\partial \gamma}{\partial x} = \frac{\partial Y}{\partial t} + v, \quad \frac{\partial \beta}{\partial x} - \frac{\partial \alpha}{\partial y} = \frac{\partial Z}{\partial t} + w, \tag{45.52}$$

$$\frac{\partial X}{\partial x} + \frac{\partial Y}{\partial y} + \frac{\partial Z}{\partial z} = \rho, \tag{45.53}$$

$$\frac{\partial \alpha}{\partial x} + \frac{\partial \beta}{\partial y} + \frac{\partial \gamma}{\partial z} = 0, \tag{45.54}$$

where the velocity of light is unity, and the Heaviside-Lorentz unit of charge is chosen so that the factor 4π disappears. The electric current u, v, w and the density of electric charge ρ form a contravariant vector, since

$$\begin{aligned}(u, v, w, \rho) &= \Sigma e\left(\frac{dx}{ds}, \frac{dy}{ds}, \frac{dz}{ds}, \frac{dt}{ds}\right) \text{ per unit volume,}^{3} \\ &= J^{\mu}, \text{ say.}\end{aligned} \tag{45.6}$$

[3]The occurrence of ds instead of dt in the denominator is due to the Michelson-Morley contraction, $\beta = dt/ds$, which makes the estimate of unit volume by a fixed observer differ from that made by an observer moving with the electrons. (Cf. equation (7.65).)

Equations (45.51) and (45.54) may be written,

$$\frac{\partial F_{\mu\nu}}{\partial x_\sigma} + \frac{\partial F_{\nu\sigma}}{\partial x_\mu} + \frac{\partial F_{\sigma\mu}}{\partial x_\nu} = 0, \tag{45.71}$$

and the remaining equations (45.52) and (45.53) give

$$\frac{\partial F^{\mu\nu}}{\partial x_\nu} = J^\mu. \tag{45.72}$$

Now (45.71) is satisfied identically on substituting the values of $F_{\mu\nu}$ from (45.2), so that (45.2) and (45.72) represent the fundamental electromagnetic equations. The former is already covariant, and the latter is made covariant by writing the covariant derivative for the ordinary derivative. Thus

$$F^{\mu\nu}_\nu = J^\mu, \tag{45.81}$$

$$F_{\mu\nu} = \frac{\partial \varkappa_\mu}{\partial x_\nu} - \frac{\partial \varkappa_\nu}{\partial x_\mu}, \tag{45.82}$$

are the required equations. These hold in the gravitational field because the conditions for the application of the principle of equivalence (§27) are satisfied.

The expression $F^{\mu\nu}_\nu$ may be simplified as in §39; but owing to the antisymmetry of $F^{\mu\nu}$ the term corresponding to (39.3) disappears, and the equation reduces to

$$\frac{1}{\sqrt{-g}}\frac{\partial}{\partial x_\nu}\left(\sqrt{-g}\,F^{\mu\nu}\right) = J^\mu. \tag{45.9}$$

The fact that Maxwell's equations can be reduced to a covariant form shows that all electromagnetic phenomena described by them will be in agreement with the principle of relativity.

46. *The Electromagnetic Energy-Tensor*

According to the electromagnetic theory, the components of mechanical force on unit volume containing electric charges are

$$\begin{aligned} k_1 &= \rho X + \gamma v - \beta w, \\ k_2 &= \rho Y + \alpha w - \gamma u, \\ k_3 &= \rho Z + \beta u - \alpha v, \end{aligned}$$

and the negative rate of doing work is

$$k_4 = -X\,u - Y\,v - Z\,w$$

since the magnetic force does no work.

By (45.41) and (45.6) these give

$$\begin{aligned} -k_\nu &= F_{\mu\nu}\,J^\mu \\ &= F_{\mu\nu}\,F^{\mu\sigma}_\sigma, \end{aligned} \tag{46.1}$$

so that k_ν is a vector.

But k_ν represents the rate at which the momentum and negative energy of the material system are being increased, i.e., in Galilean co-ordinates,

$$-\frac{\partial}{\partial x_\alpha} T_\nu^\alpha = k_\nu. \tag{46.2}$$

If there exists a corresponding tensor E_ν^α for the electromagnetic field, this must change by an equivalent amount in the opposite direction in order to satisfy the law of conservation. Thus

$$\frac{\partial}{\partial x_\alpha} E_\nu^\alpha = k_\nu. \tag{46.3}$$

It is not difficult to show from (46.1) and (46.3) that

$$E_\nu^\alpha = -F_{\nu\beta}F^{\alpha\beta} + \frac{1}{4} g_\nu^\alpha F^{\sigma\tau}F_{\sigma\tau}. \tag{46.4}$$

We omit the proof as the precise value is not of great interest to us. It is sufficient to know that the expression is of the necessary tensor-form, so that an energy-tensor for the electromagnetic field exists.

In general co-ordinates (46.2) and (46.3) are replaced by the covariant equations,

$$-T_{\nu\alpha}^\alpha = k_\nu = E_{\nu\alpha}^\alpha \tag{46.5}$$

in accordance with the principle of equivalence.

When no matter is present this gives $E_{\nu\alpha}^\alpha = 0$, and we can derive the reaction of the gravitational field just as in (39.5). It follows that electromagnetic energy in the gravitational field experiences a force just as material energy does. Further electromagnetic energy exerts gravitation, because (39.13) and (46.5) give

$$\left(G_\nu^\alpha - \frac{1}{2} g_\nu^\alpha G\right)_\alpha = 0 = -8\pi\left(T_\nu^\alpha + E_\nu^\alpha\right)_\alpha,$$

the lower α denoting covariant differentiation. Hence on integrating, (39.12) must be replaced by

$$G_\nu^\alpha - \frac{1}{2} g_\nu^\alpha G = -8\pi\left(T_\nu^\alpha + E_\nu^\alpha\right).$$

In fact the electromagnetic energy-tensor must simply be added on to the material energy-tensor throughout our work.

When $\sqrt{-g} = 1$, we have the most general law of conservation for triangular interchanges between matter, electromagnetism and gravitation.

$$\frac{\partial}{\partial x_\alpha}\left(T_\nu^\alpha + E_\nu^\alpha + t_\nu^\alpha\right) = 0. \tag{46.6}$$

47. *The Aether*

The application of the Calculus of Variations to (44.1) gives a number of differential equations equal to the number of parameters varied; but, according to a general theorem due to Hilbert, there are always four identical relations between these equations (the number 4 corresponding to the dimensions of $d\tau$). The number of independent equations is thus four less than the number of unknowns, so that in addition to arbitrary boundary conditions we can impose four arbitrary relations on the parameters. It is this freedom of choice of co-ordinates that is so fundamental a characteristic of the generalized principle of relativity

If we vary H_1 only we find the ten equations $G_{\mu\nu} = 0$. The identical relations in this case have been given in §39. If we vary the electromagnetic variable $\varkappa_\mu$ as well, we get 14 equations, of which 10 are independent, to determine 14 unknowns. Within certain limits we can give arbitrary values to four of the unknowns, and the other ten will then be determined definitely by the equations and the boundary conditions. If we elect to fix the values of the four co-ordinates $\varkappa_\mu$ in this way (so that they are, as it were, disposed of) the $g_{\mu\nu}$ will become fixed, that is to say, there will be only one possible space-time. The phenomena, electromagnetic as well as gravitational, will all be described by the $g_{\mu\nu}$ which represent the state of strain of this space-time. This space-time may be materialized as the aether, and the aether-theory does in fact attribute electromagnetic phenomena to strains in this supposed absolute medium.

This is only a crude indication of the relation of the aether-theory to our relativity theory. As is well known, the modern aether-theory involves rotational strains. Moreover, we cannot get rid of the electromagnetic variables by putting them equal to zero, because they form a vector, which cannot vanish in one system of co-ordinates without vanishing in all.

48. *Summary of the Last Two Chapters*

It may be useful to review the results which have been obtained from the point at which we introduced the energy-tensor T_μ^ν of the material system. Initially it was brought in for the practical purpose of calculating the gravitational field of a material body; but this has led on to a discussion of the general laws of dynamics.

As mentioned in §6, it is important, if we wish to adopt the principle of relativity, to show that the laws of nature which we generally accept are consistent with the principle; or if not, to modify them so that they may become consistent. We have had to modify one law – the law of gravitation. The laws of mechanics (Newton's laws of motion) are equivalent to the conservation of momentum and the conservation of mass. We have in §7(c) found it necessary to generalize the latter by admitting that energy has mass, and the conservation of mass is absorbed in the conservation of energy. The most general-statement of these two principles of conservation for material systems is found in the general equations of hydro-dynamics (or of the theory of gases), viz , (37.7) and (37.8), and it is therefore sufficient to verify these. We have done that by showing that they may be expressed in tensor-form. We have even gone further; we have shown that these laws can actually be deduced from the law of gravitation. They correspond to the four identical relations between Einstein's ten equations of gravitation (§39).

It has similarly been verified that our electromagnetic equations are of tensor-form and are therefore consistent with relativity. But in this case we have not deduced the electromagnetic equations from anything else; we have merely shown their admissibility. The energy-tensor E_μ^ν of the electromagnetic field is found from the consideration that in interchanges between the material and electromagnetic systems the total momentum and energy must remain constant.

When the co-ordinates are not Galilean, gravitational forces will be acting and the total energy and momentum of the material and electromagnetic systems will be altering. We have shown how to find this flux of energy and momentum (39.5), and in §43 we have traced it into the gravitational field, showing that it reappears there as the quantity of t_μ^ν, which, moreover, is conserved when no transfer of this kind is going on. There is, however, one reservation necessary; unlike T_μ^ν and E_μ^ν, t_μ^ν is not a tensor, and in order that this complete conservation of energy and momentum may be apparent we have to choose co-ordinates so that $\sqrt{-g} = 1$. This does not imply any exception to the physical law of conservation, because we can always choose co-ordinates satisfying this condition. It is merely that the energy-tensor is slightly more general than the physical idea of energy and momentum; the former may be reckoned with respect to any co-ordinates, the latter must be reckoned with respect to co-ordinates satisfying $\sqrt{-g} = 1$.

From the existence of an energy-tensor for the electromagnetic field, it is deduced that electromagnetic energy must experience and exert gravitational force.

The remainder of our work has been principally concerned with showing that our equations are equivalent to a principle of least action. From a theoretical standpoint there is a great deal to be said in favour of reversing the whole procedure, starting from the principle of least action as a postulate; but I have preferred the present course as more elementary.

Some difficulty may be found in the fact that the time-component of a four-dimensional vector is usually called by a different name from the space-components. The following table may be useful for reference:

Vector	*Space-Components*	*Time-Component*
$T_\mu^4 \dots$	negative momentum	energy (mass).
$T_\mu^1 \dots$	flux of negative momentum	flux of energy (mass).
$k_\mu \dots$	force	negative rate of doing work.
$\varkappa_\mu \dots$	negative vector potential	electric scalar potential.
$J_\mu \dots$	electric current-density	electric charge-density.

8 THE CURVATURE OF SPACE AND TIME

49. We have now presented the laws of gravitation, of hydromechanics, and of electromagnetism, in a form which regards all systems of co-ordinates as on an equal footing. And yet it is scarcely true to say that all systems are equally fundamental; at least we can discriminate between them in a way which the restricted principle of relativity would not tolerate.

Imagine the earth to be covered with impervious cloud. By the gyro-compass we can find two spots on it called the Poles, and by Foucault's pendulum-experiment we can determine an angular velocity about the axis through the Poles, which is usually called the earth's absolute rotation. The name "absolute rotation" may be criticized; but, at any rate, it is a name given to something which can be accurately measured. On the other hand, we fail completely in any attempt to determine a corresponding "absolute translation" of the earth. It is not a question of applying the right name there is no measured quantity to name. It is clear that the equivalence of systems of axes in relative rotation is in some way less complete than the equivalence of axes having different translations; and this may perhaps be regarded as a failure to reach the ideals of a philosophical principle of relativity.

This limitation has its practical aspect. We might suppose that from the expression (28.8) for the field of a particle at rest it would be possible by a transformation of co-ordinates to deduce the field of a particle, say, in uniform circular motion. But this is not the case. We may, of course, reduce the particle to rest by using rotating axes; but we find it necessary to take an entirely different solution of the partial differential equations, satisfying different boundary conditions.

We have not hitherto paid any attention to the invariance of the boundary conditions; and it is here that the break-down occurs. The axes ordinarily used in dynamics are such that as we recede towards infinity in space the $g_{\mu\nu}$ approach the special set of values (16.3). On transforming to other co-ordinates the differential equations are unaltered; but usually the boundary values of the $g_{\mu\nu}$, and consequently the appropriate solutions of the equations, are altered. We can, therefore, discriminate between different systems of co-ordinates according to the boundary values of the g's; and those which at infinity pass into Galilean co-ordinates may properly be considered the most fundamental, since the boundary values are most simple. The complete relativity for uniform translation is due to the boundary values as well as the differential equations remaining unaltered.[1]

We have based our theory on two axioms – the restricted principle of relativity and the principle of equivalence. These taken together may be called the *physical* principle of relativity. We have justified, or explained, them by reference to a *philosophical* principle of relativity, which asserts that experience is concerned only with the relations of objects to one another and to the observer and not to the fictitious space-time framework in which we instinctively locate them. We are now led into a dilemma; we can save this philosophical principle only by undermining its practical application. The measurement of the rotation of the earth detects something of the nature of a fundamental frame of reference – at least in the part of space accessible to observation. We shall call this the "inertial frame." Its existence does not necessarily contradict the philosophical principle, because it may, for instance, be determined by the general distribution of matter in the universe; that is to say, we may be detecting by our experiments relations to matter not generally recognized. But having recognized the existence of the inertial frame, the philosophical principle of relativity becomes arbitrary in its application. It cannot foretell that the Michelson-Morley experiment will fail to detect uniform motion relative to this frame; nor does it explain why the acceleration of the earth relative to this frame is irrelevant, but the rotation of the earth is important.

The inertial frame may be attributed (1) to unobserved world-matter, (2) to the aether, (3) to some absolute character of space-time. It is doubtful whether the discrimination between these alternatives is more than word-splitting, but they lead to rather different points of view. The last alternative seems to contradict the philosophical principle of relativity, but in the light of what has been said the physicist has no particular interest in preserving the philosophical principle. In this chapter we shall consider two suggestions towards a theory of the inertial frame made by Einstein and de Sitter respectively. These should be regarded as independent speculations, arising out of, but not required by, the theory hitherto described.

The inertial frame is distinguished by the property that the $g_{\mu\nu}$ referred to it approach the limiting Galilean values (16.3) as we recede to a great distance from all attracting matter. This is verified experimentally with considerable accuracy; but it does not follow that we can extrapolate to distances as yet unplumbed, or to infinity. If it is assumed that the Galilean values still hold at infinite distances, the inertial frame is virtually ascribed to conditions at infinity, and its explanation is removed beyond the scope of physical theory. We may, however, suppose that observational results relate to only a minute part of the whole world, and that at vaster distances the $g_{\mu\nu}$ tend to zero values which would be invariant for all finite transformations. In that case all frames of reference are alike at infinity, and the property of the inertial frame arises from conditions within a finite distance. In that case physical theories of the inertial frame may

[1]Owing to the four additional conditions that can be imposed on the g's the boundary values are not sufficient to determine the co-ordinates uniquely and the principle of relativity is valid in its most complete sense for transformations considerably more general than uniform translation.

be developed.

The ascription of the inertial frame to boundary conditions at infinity may also be avoided by abolishing the boundary. This is really only another aspect of the vanishing of the $g_{\mu\nu}$ at infinity. Our four-dimensional space-time may be regarded as a closed surface in a five-dimensional continuum; it will then be unbounded but finite, just as the surface of a sphere is unbounded.

We have seen (§44) that wherever matter exists space-time has a curvature. It might seem that if there were sufficient matter the continuum would curve round until it closed up; but it has not been found possible to eliminate the boundary so simply. I think the difficulty arises because time is not symmetrical with respect to the other co-ordinates; in general matter moves with small velocity, so that the different components of the energy-tensor T_μ^ν are not of the same order of magnitude.

50. Einstein suggests that in measurements on a vast scale the line-element has the form

$$ds^2 = -R^2\{d\chi^2 + \sin^2\chi\,(d\theta^2 + \sin^2\theta\,d\varphi^2)\} + dt^2. \tag{50.1}$$

This expression includes the effects of the general distribution of matter through space; but there will be superposed the local irregularities due to its condensation into stellar systems, etc.

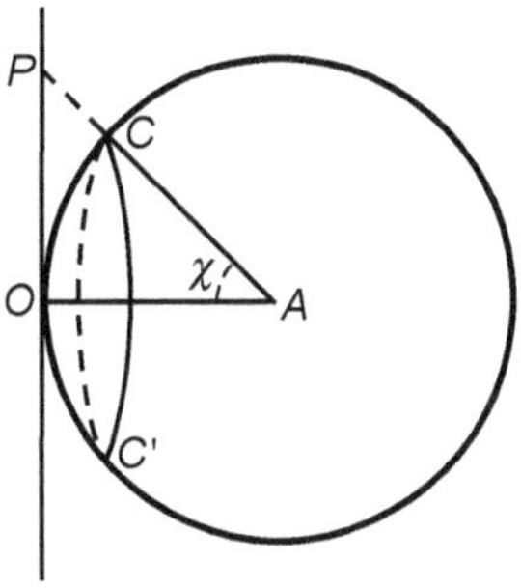

Fig. 5

The expression (50.1) can be interpreted[2] as belonging to a three-dimensional space which forms the surface of a hyper-sphere of radius R in four dimensions, the time being rectilinear. Let O be the origin of co-ordinates (Fig. 5), A the center of the hydrosphere, and χ the angle OAC. If θ is the azimuthal angle of the plane OAC, the line-element at C for an ordinary sphere would be

$$-ds^2 = R^2\,d\chi^2 + R^2\,\sin^2\chi\,d\theta^2.$$

The expression (50.1) is the extension of this for an extra dimension measured by φ.

In the figure the circumference of the circle CC' is $2\pi R\sin\chi$, but its radius measured along the sphere is R_χ. Similarly in our curved space the surface of

[2]Other interpretations are possible; but this is probably the easiest conception for those unfamiliar with non-Euclidean geometry. For this reason I do not here describe the interpretation in terms of "elliptical space," which has certain advantages.

a sphere of radius R_χ will be $4\pi R^2 \sin^2\chi$ successively more distant spheres will increase in area up to a radius $\frac{1}{2}\pi R$, and afterwards decrease to a point for the limiting distance πR. The whole volume of space is finite and equal to $2\pi^2 R^3$ in natural measure.[3]

From (50.1) the values of $G_{\mu\nu}$ can be calculated just as in §28. We find, in fact,

$$\begin{aligned} G_{\mu\nu} &= \frac{2}{R^2}\, g_{\nu\nu}, \quad \text{except } G_{44} = 0 \\ \text{so that} \qquad G &= \frac{6}{R^2}. \end{aligned} \tag{50.2}$$

Hence by (35.8)

$$-\,8\pi\, T_{\mu\nu} = -\frac{1}{R^2}\, g_{\mu\nu}, \text{ except } \; -\,8\pi\, T_{44} = -\frac{3}{R^2}. \tag{50.3}$$

Unless we are willing to suppose that the matter in the universe is moving with speeds approaching that of light, T_{44} is much greater than the other components, and it is clearly impossible to satisfy (50.3). The only possible course is to make a slight modification of the law of gravitation. Neglecting the motion of matter we shall have $T_{44} = \rho$, and the other components vanish. The modified law that satisfies (50.2) must then be

$$G_{\mu\nu} - \frac{1}{2} g_{\mu\nu}\,(G - 2\lambda) = -8\pi\, T_{\mu\nu}, \tag{50.4}$$

where $\lambda = 1/R^2$ and $\rho = 1/4\pi R^2$.

Equation (50.4) replaces (35.8). The radius R may be as great as we please, so that we may satisfy our scruples without introducing any modification perceptible to observation.

In Hamilton's principle G becomes replaced by $G - 4\lambda$, and space-time has a natural curvature 4λ when no matter is present; this curvature is increased to 6λ where there is matter having the average density. (Cf. (44.4) with $k_4 = 0$.)

Since the whole volume of space in natural measure is $2\pi^2 R^3$ the total mass of matter is $2\pi^2 R^3 \rho = \frac{1}{2}\pi R$. The mass of the sun is 1.47 kilometers; the mass of the stellar system may be estimated at $10^9 \times$ sun; let us suppose further that the spiral nebulae represent 1,000,000 stellar systems having this mass. Even this total mass will only give us a universe of radius 10^{15} kilometers, or about 30 parsecs much less than the average distance of the naked-eye stars. Einstein's hypothesis therefore demands the existence of vast quantities of undetected matter which we may call world-matter.

Some curious results are obtained by following out the properties of this spherical space. The parallax of a star diminishes to zero as the distance (in natural measure) increases up to $\frac{1}{2}\pi R$; it then becomes negative and reaches $-\,90^0$ at a distance φR. Apart from absorption of light in space we should see an anti-sun, at the point of the sky opposite to the sun equally large and equally bright,[4] the surface-markings corresponding to the back of the sun. After

[3] The observer will probably introduce measures more convenient to himself (cf. §52), so that in his co-ordinates the limiting distance may be ∞ or even beyond.

[4] Disregarding the sun's absolute motion referred to later.

travelling "round the world" the sun's rays come back to a focus. Since ρ and R are related, it has been suggested that we can use the invisibility of this anti-sun to give a lower limit to R, assuming that no light is lost in space except by the scattering action of the world-matter. But it appears to have been overlooked that Einstein's new hypothesis is inconsistent with relativity in its ordinary sense; the anti-sun will not be a virtual image of the sun as it is now, but of the sun as it was when it emitted the light – perhaps millions of years ago, when it was in another part of the stellar system. Einstein has restored the differentiation between space and time by assuming the space-time world to be cylindrical, so that the linear direction gives an absolute time. It is only locally that we can still make Minkowski's transformation; rigorously the physical principle of relativity is violated since space-time is no longer isotropic.

We regret being unable to recommend this rather picturesque theory of anti-suns and anti-stars. It suggests that only a certain proportion of the visible stars are material bodies, the remainder are ghosts of stars, haunting the places where stars used to be in a far-off past.

Owing to this violation of the restricted principle of relativity we have a feeling that Einstein's new hypothesis throws away the substance for the shadow. It is also open to the serious criticism that the law of gravitation is made to involve a constant λ, which depends on the total amount of matter in the universe ($\lambda = \pi^2/4M^2$). This seems scarcely conceivable; and it looks as though the solution involves a very artificial adjustment.

51. An alternative proposal has been made by de Sitter which seems much less open to objection. He takes for the line element

$$ds^2 = -R^2\{d\chi^2 + \sin^2\chi\,(d\theta^2 + \sin^2\theta\,d\varphi^2)\} + \cos^2\chi\,dt^2. \tag{51.1}$$

For constant time the three-dimensional space is spherical as in (50.1); but there is also a curvature in the time-variable.

With the present variables this is not of a simple kind, but setting

$$\begin{aligned} \sin\chi &= \sin\zeta\,\sin\omega \\ \tan\left(\frac{it}{R}\right) &= \cos\zeta\,\tan\omega \end{aligned} \tag{51.2}$$

we find

$$ds^2 = -R^2\big(d\omega^2 + \sin^2\omega(d\zeta^2 + \sin^2\zeta\,(d\theta^2 + \sin^2\theta\,d\varphi^2))\big) \tag{51.3}$$

which corresponds to spherical polar co-ordinates $(R, \omega, \zeta, \theta, \varphi)$ in space of five dimensions. By measuring ζ from different azimuths we perform an operation corresponding to Minkowski's rotation of the time-axis, so that there is here no absolute time, and the original principle of relativity is fully satisfied.

The properties of de Sitter's space-time are best recognized from (51.1). Near the origin we have ordinary Galilean space- time. As we recede, space has the spherical properties already mentioned, and in addition measured time (ds) begins to run slow relative to co-ordinate time (dt). Finally at $\chi = \frac{1}{2}\pi$ i.e., at a natural distance $\frac{r}{2}\pi R$, time stands still. At any fixed point ds is zero however large dt may be, so that nothing whatever can happen however long we wait.

Of course, this is merely the point of view of the observer at the origin of co-ordinates. All parts of this spherical continuum are interchangeable; and if our observer could transport himself to this peaceful abode, he would find Nature there as active as ever. Moreover, adopting the co-ordinates natural to his new position, he would judge his old home to be in this passive state. There is a complete lack of correspondence between the times at the two places. They are, as it were, at right angles, so that the progress of time at one point has no relation to the perception of time at the other point.

The line-element (51.1) leads to

$$G_{\mu\nu} = \frac{3}{R^2} g_{\mu\nu}$$

and accordingly the law of gravitation is taken to be (50.4), with

$$\lambda = \frac{3}{R^2}.$$

The aggregate curvature due to matter is here neglected in comparison with the natural curvature due to the modification of the law of gravitation, and there is no assumption of the existence of vast quantities of matter not yet recognized.

There is no anti-sun on de Sitter's hypothesis, because light, like everything else, is reduced to rest at the zone where time stands still, and it can never get round the world. The region beyond the distance $\frac{1}{2}\pi R$ is altogether shut off from us by this barrier of time. The parallax of a star at this distance will be such as corresponds to a distance R in Euclidean space, and this is the minimum value possible.

The most interesting application of this hypothesis is in connection with the very large observed velocities of spiral nebulae, which are believed to be distant sidereal systems. Since $\sqrt{g_{44}} = \cos\chi$ the vibrations of the atoms become slower (in the observer's time) as $\cos\chi$ diminishes, in accordance with §34. We should thus expect the spectral lines to be displaced towards the red in very distant objects, an effect which would in practice be attributed to a great velocity of recession. It is not possible to say as yet whether the spiral nebulae show a systematic recession, but so far as determined up to the present receding nebulae seem to preponderate.

Superposed on the (spurious) systematic radial velocity will be the individual velocities of the nebulae. It is scarcely possible to say what these are likely to be without making some assumption. There is no meaning in absolute motion, and if two systems are entirely independent, so that their relative motion has no physical cause, it must be quite arbitrary, and there is no reason to expect it to be small compared with the velocity of light. If, however, the systems have separated from one another, it can be shown by rather laborious calculations[5] that their velocities will tend to become more diverse as they recede, up to the

[5]De Sitter, "Monthly Notices," November, 1917.

limit $\frac{1}{4}\pi R$ for which the velocities are comparable with that of light. We should thus have an explanation of the large velocities of the spirals, averaging 300-400 km. per sec., and we could perhaps form an estimate of the value of R.

It must be remembered that in natural measure the internal motions of stars in a spiral system will be of the same magnitude as in our own system, owing to the homogeneous character of de Sitter's space-time. In co-ordinate measure these internal motions will be smaller owing to the transformation of the time. The possibility of large divergent motions of the systems as a whole depends on the large separation between them.

52. So far we have used spherical co-ordinates, but we can map the spherical space of Einstein or of de Sitter on a fiat space by performing the central projection $r = R\tan\chi$, r will be represented by OP in Fig. 5, and the variables r, θ, φ will satisfy Euclidean geometry. This does not mean that measured space is Euclidean; but that we multiply our measures by suitable factors in order to obtain results which will fit together in Euclidean space, just as we did for a local gravitational field in §28. With r as variable (50.1) and (51.1) become, respectively,

$$ds^2 = \frac{-dr^2}{(1+\varepsilon r^2)^2} - \frac{r^2}{1+\varepsilon r^2}(d\theta + \sin^2\theta\, d\varphi^2) + dt^2. \tag{52.1}$$

$$ds^2 = \frac{-dr^2}{(1+\varepsilon r^2)^2} - \frac{r^2}{1+\varepsilon r^2}(d\theta^2 + \sin^2\theta\, d\varphi^2) + \frac{dt^2}{1+\varepsilon r^2}, \tag{52.2}$$

where $\varepsilon = 1/R^2$

These show that at "infinity" (i.e., $r = \infty$) the values of $g_{\mu\nu}$ in rectangular co-ordinates approach the respective limits.

Einstein				de Sitter				Galileo			
0	0	0	0	0	0	0	0	-1	0	0	0
0	0	0	0	0	0	0	0	0	-1	0	0
0	0	0	0	0	0	0	0	0	0	-1	0
0	0	0	1	0	0	0	0	0	0	0	1

the Galilean values being added for comparison.

De Sitter's limiting values are invariant for all transformations; Einstein's only for transformations not involving the time; the Galilean values for the transformation of uniform motion and a limited group of other transformations.

De Sitter's hypothesis thus appears to present the greatest advantages; but it will not satisfy the followers of Mach's philosophy. He derives his inertial frame from the spherical property of space-time which in turn is derived from the slightly modified law of gravitation; it is not determined by anything material. The followers of Mach maintain that if there were no matter there could be no inertial frame, and it appears that this is Einstein's reason for preferring his own suggestion. In his theory if all matter were abolished, R would become zero and the world would vanish to a point. There is something rather fascinating in a theory of space by which, the more matter there is, the more room is provided. It is satisfactory, too, from Einstein's standpoint, because he is unwilling to admit that a thinkable space without matter could exist. For our part, we feel equally unwilling to assent to the introduction of vast quantities of world-matter, which, (to quote de Sitter) "fulfils no other purpose than to enable us to suppose it not to exist."

53. In this discussion of the law of gravitation, we have not sought, and we have not reached, any ultimate explanation of its cause. A certain connection between the gravitational field and the measurement of space has been postulated, but this throws light rather on the nature of our measurements than on gravitation itself. The relativity theory is indifferent to hypotheses as to the nature of gravitation, just as it is indifferent to hypotheses as to matter and light. We do not in these days seek to explain the behaviour of natural forces in terms of a mechanical model having the familiar characteristics of matter in bulk; we have to accept some mathematical expression as an axiomatic property which cannot be further analyzed. But I do not think we have reached this stage in the case of gravitation.

There are three fundamental constants of nature which stand out pre-eminently:

The velocity of light	3.00×10^{10}	C.G.S. units	dimensions LT^{-1}
The quantum	6.55×10^{-27}	"	" ML^2T^{-1}
The constant of gravitation	6.66×10^{-8}	"	" $M^{-1}L^3T^{-2}$

From these we can construct a fundamental unit of length whose value is

$$4 \times 10^{-33}\text{cms.}$$

There are other natural units of length – the radii of the positive and negative unit electric charges – but these are of an altogether higher order of magnitude.

With the possible exception of Osborne Reynolds's theory of matter, no theory has attempted to reach such fine-grainedness. But it is evident that this length must be the key to some essential structure. It may not be an unattainable hope that some day a clearer knowledge of the processes of gravitation may be reached; and the extreme generality and detachment of the relativity theory may be illuminated by the particular study of a precise mechanism.

APPENDIX

PHILOSOPHICAL TRANSACTIONS OF THE ROYAL SOCIETY A
MATHEMATICAL, PHYSICAL & ENGINEERING SCIENCES

A Determination of the Deflection of Light by the Sun's Gravitational Field, from Observations Made at the Total Eclipse of May 29, 1919

F. W. Dyson, A. S. Eddington and C. Davidson

Phil. Trans. R. Soc. Lond. A 1920 **220**, doi: 10.1098/rsta.1920.0009, published 1 January 1920

IX. *A Determination of the Deflection of Light by the Sun's Gravitational Field, from Observations made at the Total Eclipse of May* 29, 1919.

By Sir F. W. Dyson, *F.R.S., Astronomer Royal, Prof.* A. S. Eddington, *F.R.S., and Mr.* C. Davidson.

(*Communicated by the Joint Permanent Eclipse Committee.*)

Received October 30,—Read November 6, 1919.

[Plate 1.]

Contents.

I. Purpose of the Expeditions.

1. The purpose of the expeditions was to determine what effect, if any, is produced by a gravitational field on the path of a ray of light traversing it. Apart from possible surprises, there appeared to be three alternatives, which it was especially desired to discriminate between—

(1) The path is uninfluenced by gravitation.

(2) The energy or mass of light is subject to gravitation in the same way as ordinary matter. If the law of gravitation is strictly the Newtonian law, this leads to an apparent displacement of a star close to the sun's limb amounting to $0''\cdot 87$ outwards.

(3) The course of a ray of light is in accordance with Einstein's generalised relativity theory. This leads to an apparent displacement of a star at the limb amounting to $1''\cdot 75$ outwards.

In either of the last two cases the displacement is inversely proportional to the distance of the star from the sun's centre, the displacement under (3) being just double the displacement under (2).

It may be noted that both (2) and (3) agree in supposing that light is subject to gravitation in precisely the same way as ordinary matter. The difference is that, whereas (2) assumes the Newtonian law, (3) assumes Einstein's new law of gravitation. The slight

 [Published April 27, 1920.

deviation from the Newtonian law, which on EINSTEIN's theory causes an excess motion of perihelion of Mercury, becomes magnified as the speed increases, until for the limiting velocity of light it doubles the curvature of the path.

2. The displacement (2) was first suggested by Prof. EINSTEIN* in 1911, his argument being based on the Principle of Equivalence, viz., that a gravitational field is indistinguishable from a spurious field of force produced by an acceleration of the axes of reference. But apart from the validity of the general Principle of Equivalence there were reasons for expecting that the electromagnetic energy of a beam of light would be subject to gravitation, especially when it was proved that the energy of radio-activity contained in uranium was subject to gravitation. In 1915, however, EINSTEIN found that the general Principle of Equivalence necessitates a modification of the Newtonian law of gravitation, and that the new law leads to the displacement (3).

3. The only opportunity of observing these possible deflections is afforded by a ray of light from a star passing near the sun. (The maximum deflection by Jupiter is only 0″·017.) Evidently, the observation must be made during a total eclipse of the sun.

Immediately after EINSTEIN's first suggestion, the matter was taken up by Dr. E. FREUNDLICH, who attempted to collect information from eclipse plates already taken; but he did not secure sufficient material. At ensuing eclipses plans were made by various observers for testing the effect, but they failed through cloud or other causes. After EINSTEIN's second suggestion had appeared, the Lick Observatory expedition attempted to observe the effect at the eclipse of 1918. The final results are not yet published. Some account of a preliminary discussion has been given,† but the eclipse was an unfavourable one, and from the information published the probable accidental error is large, so that the accuracy is insufficient to discriminate between the three alternatives.

4. The results of the observations here described appear to point quite definitely to the third alternative, and confirm EINSTEIN's generalised relativity theory. As is well-known the theory is also confirmed by the motion of the perihelion of Mercury, which exceeds the Newtonian value by 43″ per century—an amount practically identical with that deduced from EINSTEIN's theory. On the other hand, his theory predicts a displacement to the red of the Fraunhofer lines on the sun amounting to about 0·008 Å in the violet. According to Dr. ST. JOHN‡ this displacement is not confirmed. If this disagreement is to be taken as final it necessitates considerable modifications of EINSTEIN's theory, which it is outside our province to discuss. But, whether or not changes are needed in other parts of the theory, it appears now to be established that EINSTEIN's law of gravitation gives the true deviations from the Newtonian law both for the relatively slow-moving planet Mercury and for the fast-moving waves of light.

It seems clear that the effect here found must be attributed to the sun's gravitational field and not, for example, to refraction by coronal matter. In order to produce the

* 'Annalen der Physik,' vol. XXXV, p. 898.

† 'Observatory,' vol. XLII, p. 298.

‡ 'Astrophysical Journal,' vol. XLVI, p. 249.

observed effect by refraction, the sun must be surrounded by material of refractive index $1 + \cdot 00000414/r$, where r is the distance from the centre in terms of the sun's radius. At a height of one radius above the surface the necessary refractive index $1 \cdot 00000212$ corresponds to that of air at $\frac{1}{140}$ atmosphere, hydrogen at $\frac{1}{60}$ atmosphere, or helium at $\frac{1}{20}$ atmospheric pressure. Clearly a density of this order is out of the question.

II. Preparations for the Expeditions.

5. In March, 1917,* it was pointed out as the result of an examination of the photographs taken with the Greenwich astrographic telescope at the eclipse of 1905 that this instrument was suitable for the photography of the field of stars surrounding the sun in a total eclipse. Attention was also drawn to the importance of observing the eclipse of May 29, 1919, as this afforded a specially favourable opportunity owing to the unusual number of bright stars in the field, such as would not occur again for many years.

With weather conditions as good as those at Sfax in the 1905 eclipse—and these were by no means perfect—it was anticipated that twelve stars would be shown. Their positions are indicated in the diagram on next page, on which is also marked on the same scale the outline of a 16 × 16 cm. plate (used with the astrographic telescopes of 3·43 metres focal length) and a 10 × 8-inch plate (used with a 4-inch lens of 19 feet focal length).

The following table gives the photographic magnitudes and standard co-ordinates of the stars, and the gravitational displacements in x and y calculated on the assumption of a radial displacement $1'' \cdot 75 \frac{r_0}{r}$, where r is the distance from the sun's centre and r_0 the radius of the sun.

Table I.

No.	Names.	Photog. Mag.	Co-ordinates. Unit = 50′.		Gravitational displacement.			
					Sobral.		Principe.	
			x.	y.	x.	y.	x.	y.
		m.			″	″	″	″
1	B.D., 21°, 641	7·0	+0·026	−0·200	−1·31	+0·20	−1·04	+0·09
2	Piazzi, IV, 82	5·8	+1·079	−0·328	+0·85	−0·09	+1·02	−0·16
3	κ^2 Tauri	5·5	+0·348	+0·360	−0·12	+0·87	−0·28	+0·81
4	κ^1 Tauri	4·5	+0·334	+0·472	−0·10	+0·73	−0·21	+0·70
5	Piazzi, IV, 61	6·0	−0·160	−1·107	−0·31	−0·43	−0·31	−0·38
6	υ Tauri	4·5	+0·587	+1·099	+0·04	+0·40	+0·01	+0·41
7	B.D., 20°, 741	7·0	−0·707	−0·864	−0·38	−0·20	−0·35	−0·17
8	B.D., 20°, 740	7·0	−0·727	−1·040	−0·33	−0·22	−0·29	−0·20
9	Piazzi, IV, 53	7·0	−0·483	−1·303	−0·26	−0·30	−0·26	−0·27
10	72 Tauri	5·5	+0·860	+1·321	+0·09	+0·32	+0·07	+0·34
11	66 Tauri	5·5	−1·261	−0·160	−0·32	+0·02	−0·30	+0·01
12	53 Tauri	5·5	−1·311	−0·918	−0·28	−0·10	−0·26	−0·09
13	B.D., 22°, 688	8·0	+0·089	+1·007	−0·17	+0·40	−0·14	+0·39

* 'Monthly Notices, R.A.S.,' LXXVII, p. 445.

It may be noted that No. 1 is lost in the corona on the photographs taken at Sobral. The star, No. 13, of magnitude 8·0, is shown on some of the astrographic plates at Sobral.

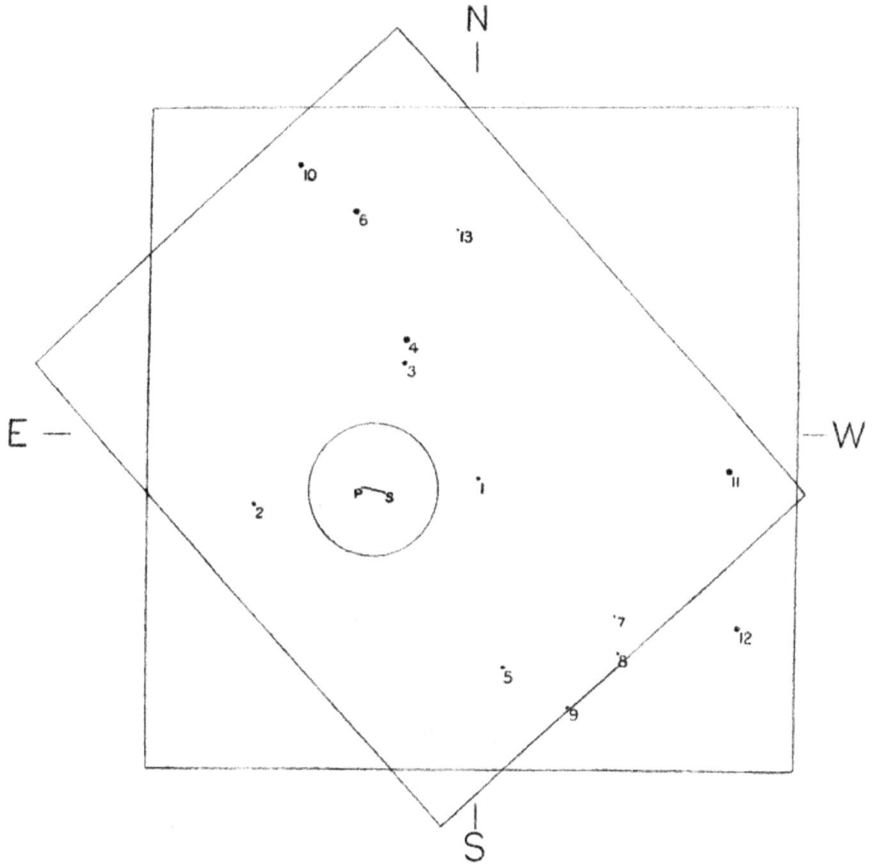

Diagram 1.

6. The track of the eclipse runs from North Brazil across the Atlantic, skirting the African coast near Cape Palmas, passing through the Island of Principe, then across Africa to the western shores of Lake Tanganyika. Enquiry as to the suitable sites and probable weather conditions was kindly made by Mr. Hinks. It appeared that a station in North Brazil, the Island of Principe, and a station on the west of Lake Tanganyika were possible. A station near Cape Palmas did not seem desirable from the meteorological reports though, as the event proved, the eclipse was observed in a cloudless sky

by Prof. BAUER, who was there on an expedition to observe magnetic effects. At the station at Tanganyika it was thought the sun was at too low an altitude for observations of this character, owing to the large displacements which would be caused by refraction.

A circular received from Dr. MORIZE, the director of the Observatory at Rio, stated that Sobral was the most suitable station in North Brazil and gave copious information of the meteorological conditions, mode of access, &c.

7. Acting on this information the Joint Permanent Eclipse Committee at a meeting on November 10, 1917, decided, if possible, to send expeditions to Sobral in North Brazil, and to the island of Principe. Application was made to the Government Grant Committee for £100 for instruments and £1,000 for the expedition, and a sub-committee consisting of Sir F. W. DYSON, Prof. EDDINGTON, Prof. FOWLER and Prof. TURNER was appointed to make arrangements for the expeditions. This sub-committee met in May and June, 1918, and made provisional arrangements for Prof. EDDINGTON and Mr. COTTINGHAM to take the object glass of the Oxford astrographic telescope to Principe, and Mr. DAVIDSON and Father CORTIE to take the object glass of the Greenwich astrographic telescope to Sobral. It was arranged for the clocks and mechanism of the cœlostats to be overhauled by Mr. COTTINGHAM. Preliminary inquiries were also set on foot as to shipping facilities, from which it appeared very doubtful whether the expeditions could be carried through.

Conditions had changed materially in November, 1918, and at a meeting of the sub-committee on November 8, it was arranged to assemble the instruments at Greenwich, and make necessary arrangements with all speed for the observers to leave England by the end of February, 1919. In addition to the astrographic object glasses fed by 16-inch cœlostats, Father CORTIE suggested to the sub-committee the use of the 4-inch telescope of 19-feet focus, which he had used at Hernosand, Sweden, in 1914, in conjunction with an 8-inch cœlostat, the property of the Royal Irish Academy. It was arranged to ask for the loan of these instruments. As Father CORTIE found it impossible to spare the necessary time for the expedition his place was taken by Dr. CROMMELIN of the Royal Observatory.

8. In November, 1918, the only workman available at the Royal Observatory was the mechanic, the carpenter not having been released from military service. In these circumstances Mr. BOWEN, the civil engineer at the Royal Naval College, was consulted. He kindly undertook the construction of frame huts covered with canvas, which could be easily packed and readily put together. These were generally similar to those used in previous expeditions from the Royal Observatory (see 'Monthly Notices,' Vol. LVII., p. 101). He also lent the services of a joiner who worked at the Observatory on the woodwork of the instruments.

It was found possible to obtain steel tubes for the astrographic objectives. These were, for convenience of carriage, made in two sections which could be bolted together. The tubes were provided with flanges at each end, the objective being attached to one of these, and a wooden breech piece to the other. In the breech piece suitable provision was made for the focussing and squaring on of the plates. The plate holders were of a simple construction, permitting the plate to be pushed into contact with three metal tilting screws on the breech piece thus insuring a constancy of focal plane. Eighteen plate-carriers were obtained for each of the astrographic telescopes, made according to a pattern supplied.

With the 4-inch lens Father CORTIE lent the square wooden tube used by him in 1914. This was modified at the breech end to secure greater rigidity and constancy of focus.

It was designed for dark slides carrying 10 × 8 inch plates, and four of these, carrying eight plates, were lent with the telescope. The desirability of using larger plates was considered, but the time at disposal to make the necessary alterations was insufficient.

The 16-inch cœlostats which had been overhauled by Mr. COTTINGHAM were mounted and tested as far as the unfavourable weather conditions of February, 1919, would permit. The 8-inch cœlostat was constructed for these latitudes. To make it serviceable near the equator a strong wooden wedge was made on which the cœlostat was bolted.

The 8-inch mirror was silvered at the observatory, but owing to lack of facilities for maintaining a uniform temperature approaching 60° F. in the wintry weather of February, the larger mirrors were sent away to be silvered.

Photographic plates, suitably packed in hermetically sealed tin boxes, were obtained from the Ilford and Imperial Companies. The Ilford plates employed were Special Rapid and Empress, and those of the Imperial Company, Special Sensitive, Sovereign and Ordinary.

The instruments were carefully packed and sent to Liverpool a week in advance, with the exception of the objectives. These were packed in cases inside hampers and remained under the personal care of the observers, who embarked on the "Anselm" on March 8.

III. THE EXPEDITION TO SOBRAL.

(*Observers, Dr.* A. C. D. CROMMELIN *and Mr.* C. DAVIDSON.)

9. Sobral is the second town of the State of Ceara, in the north of Brazil. Its geographical co-ordinates are: longitude 2h. 47m. 25s. west; latitude 3° 41′ 33″ south; altitude 230 feet. Its climate is dry and though hot not unhealthy.

The expedition reached Para on the "Anselm" on March 23. There was a choice of proceeding immediately to Sobral or waiting for some weeks. It was considered undesirable to go there before we heard from Dr. MORIZE what arrangements were being made, so we reported our arrival to him by telegram and decided to await his reply. As we had thus some time on our hands we continued the voyage to Manaos in the "Anselm," returning to Para on April 8.

By the courtesy of the Brazilian Government our heavy baggage was passed through the customs without examination and we continued our journey to Sobral, leaving Para on April 24 by the steamer "Fortaleza" and arriving at Camocim on April 29.

Here we were met by Mr. JOHN NICOLAU, who had been instructed to assist us with our baggage through to Sobral. We proceeded from Camocim to Sobral by train on April 30, our baggage following the next day.

We were met at Sobral station by representatives of both the Civil and Ecclesiastical Authorities, headed respectively by Dr. JACOME D'OLIVEIRA, the Prefect, and Mgr. FERREIRA, and conducted to the house which had been placed at our disposal by the kindness of its owner, Col. VICENTE SABOYA, the Deputy for Sobral. We were joined there nine days later by the Washington (Carnegie) Eclipse Commission, consisting of Messrs. DANIEL WISE and ANDREW THOMSON.

We are greatly indebted to Dr. LEOCADIO ARAUJO, of the State Ministry of Agriculture, who had been deputed to interpret for us and to assist us in our preparations. His services were invaluable, and contributed greatly to our success, as also to our well-being during our stay.

10. A convenient site for the eclipse station offered itself just in front of the house; this was the race-course of the Jockey Club, and was provided with a covered grand stand, which we found most convenient for unpacking and storage and in the preparatory work. We laid down a meridian line, after which brick piers were constructed for the cœlostats and for the steel tube of the astrographic telescope. Whilst this was in progress the huts were being erected.

The pier of the small cœlostat was constructed so as to leave a clear space in the middle of one end for the fall of the weight, which was thus below the driving barrel of the clock. By continuing the hole below the foundations of the pier, space was provided for a fall of the weight permitting a run of 25 minutes. In the case of the 16-inch cœlostat, the clock was mounted on the top of a long wooden trunk, nearly 4 feet in length, which was placed on end, and sunk in the earth to a depth of about 2 feet. The weight descended inside the trunk directly from the driving barrel, and had space for a continuous run of over half-an-hour.

The 16-inch cœlostat had free adjustment for all latitudes; but the 8-inch one, constructed for European latitudes, was mounted on a wooden base, inclined at an angle of about 40 degrees, constructed before leaving Greenwich. The clock had to be separated from the cœlostat, mounted on a wooden base and reversed, to adjust to the Southern Hemisphere. It performed very satisfactorily, and no elongation of the star images is shown with 28 seconds' exposure.

To provide for the changing declination of the sun the piers of the astrographic telescope were made with grooves in the top, in which the wooden V-supports of the tube could slide, thus allowing for the change of azimuth.

The tube of the astrographic telescope was circular in section, and could rest in any position in the Vs; for convenience it was adjusted so that the directions of R.A. and declination were parallel to the sides of the plate; this involved a tilt of the plate holders of about 4 degrees to the horizontal.

The 4-inch lens was taken as an auxiliary; we used the square wooden tube, 19 feet in length, originally used by Father CORTIE at Hernosand in 1914, together with the 10 × 8-inch plate carriers. Study of the star-diagram showed that seven stars could be photographed by turning the plate through 45 degrees. The tube was therefore placed on its angle, large wooden V-supports being prepared to fit the tube; these rested on strong wooden trestles.

The focussing was at first done visually on Arcturus, using an eyepiece fitted with a cobalt glass (after the plate supports and object-glass had been adjusted for perpendicularity to the axis). A series of exposures was then made, the focus being varied slightly so as to cover a sufficient range. Examination of these photographs showed at once that there was serious astigmatism due to the figure of the mirror of the 16-inch cœlostat. By inserting an 8-inch stop this was reduced to a large extent, and this stop was henceforth used throughout; but the defect was of such a character that it was clear that it would be necessary to stay at Sobral and obtain comparison plates of the eclipse field in July when the sun had moved away.

The focus of the 4-inch was determined in a similar manner. The images, though superior to those of the astrographic, were not quite perfect, and here again comparison plates in July were necessary. Once the focus had been decided on, the breech end was securely screwed up to avoid any chance of subsequent movement.

A few check plates of the field near Arcturus were taken, but have not been used.

11. The following is a summary of the meteorological conditions during our stay. The barometer record was interesting in that it showed very little change from day to day, in spite of changes in the type of weather; there was, however, a very well marked semi-diurnal variation, with range of about 0·15 inch. The temperature range was fairly uniform, from a maximum of about 97° F. towards 3 p.m. to a minimum of about 75° F. at 5 a.m. The relative humidity (as shown by a hygrograph belonging to the Brazilian Commission) followed the temperature closely, varying from 30 per cent. in the afternoon to 90 per cent. in the early morning.

May is normally the last month of the rainy season at Sobral, but this year the rainfall was very scanty; there were a few afternoon showers, each ushered in by a violent gust of wind; and on May 25 there was very heavy rain, which was welcome for its moistening effect on the ground, the dust hitherto having been troublesome to the clockwork although every care had been taken to protect it. There was a fair amount of cloud in the mornings, but the afternoons and nights were clear in the majority of cases. Mt. Meruoca, 2,700 feet high, about 6 miles to the N.W., was a collector of cloud, its summit being frequently veiled in mist. In spite of its cooler climate, the summit would thus not have been a suitable eclipse station, and, in fact, nothing of the total phase of the eclipse was seen from it.

12. Although water was generally scarce, we were very fortunately situated as we enjoyed an unlimited supply of good water laid on at the house. This was of great benefit in the photographic operations. Ice was unobtainable, but by the use of earthenware water-coolers it was possible to reduce the temperature to about 75°, and by working

only at night or before dawn development of the plates was fairly easy. Formalin was used in every case to harden the films, and thereby minimise the chance of distortion due to the softening of the films by the warm solutions.

We had provided ourselves with two brands of plates, but it had become apparent from photographs taken and developed before the eclipse that one of these brands was unsuitable in the hot climate, and it was decided to use practically only one brand of plates.

In taking the experimental photographs it was noticed that the clocks and cœlostats were very sensitive to wind. We had reason to fear strong gusts about the time of totality, such as had occurred in other eclipses; and as the conditions of our locality seemed to render them specially probable, protective wind screens were erected round the hut openings at every point where it was possible without interfering with the field of view. Happily dead calm prevailed at the critical time. Screens also protected all projecting parts of the telescope tubes from direct sunlight.

The performance of the 16-inch cœlostat was unsatisfactory in respect of driving. There was a clearly marked oscillation of the images on the screen in a period of about 30 seconds. For this reason exposure time was shortened, so as to multiply the number of exposures in the hope that some would be near the stationary points.

13. The morning of the eclipse day was rather more cloudy than the average, and the proportion of cloud was estimated at $\frac{9}{10}$ at the time of first contact, when the sun was invisible; it appeared a few seconds later showing a very small encroachment of the moon, and there were various short intervals of sunshine during the partial phase which enabled us to place the sun's image at its assigned position on the ground glass, and to give a final adjustment to the rates of the driving clocks. As totality approached, the proportion of cloud diminished, and a large clear space reached the sun about one minute before second contact. Warnings were given 58s., 22s. and 12s. before second contact by observing the length of the disappearing crescent on the ground glass. When the crescent disappeared the word "go" was called and a metronome was started by Dr. Leocadio, who called out every tenth beat during totality, and the exposure times were recorded in terms of these beats. It beat 320 times in 310 seconds; allowance has been made for this rate in the recorded times. The programme arranged was carried out successfully, 19 plates being exposed in the astrographic telescope with alternate exposures of 5 and 10 seconds, and eight in the 4-inch camera with a uniform exposure of 28 seconds. The region round the sun was free from cloud, except for an interval of about a minute near the middle of totality when it was veiled by thin cloud, which prevented the photography of stars, though the inner corona remained visible to the eye and the plates exposed at this time show it and the large prominence excellently defined. The plates remained in their holders until development, which was carried out in convenient batches during the night hours of the following days, being completed by June 5.

14. No observation of contact times was made, but it is known that these times were

somewhat before those calculated. As the times recorded were reckoned from second contact, it is assumed that this occurred May 28, 23h. 58m. 18s. G.M.T.

The details of the exposures are given in the following tables :—

EXPOSURES with the 13-inch Astrographic Telescope stopped to 8 inches.

Ref. No.	G.M.T. at Commencement of Exposure.	Exposure.	Plate.	Ref. No.	G.M.T. at Commencement of Exposure.	Exposure.	Plate.
	d. h. m. s.	s.			d. h. m. s.	s.	
1	28 23 58 23	5	O.	11	29 0 1 7	5	S.R.
2	37	10	E.	12	22	10	E.
3	57	5	E.	13	36	5	E.
4	59 11	10	S.	14	51	10	S.R.
5	30	5	S.R.	15	2 10	5	S.R.
6	45	10	S.R.	16	25	10	S.R.
7	29 0 0 4	5	S.R.	17	44	5	E.
8	19	10	E.	18	58	10	E.
9	39	5	E.	19	3 18	5	O.
10	53	10	S.R.				

EXPOSURES with the 4-inch Telescope.

Ref. No.	G.M.T. at Commencement of Exposure.	Exposure.	Plate.	Ref. No.	G.M.T. at Commencement of Exposure.	Exposure.	Plate.
	d. h. m. s.	s.			d. h. m. s.	s.	
1	28 23 58 21	28	S.R.	5	29 0 0 56	28	S.R.
2	59 0	28	S.R.	6	1 34	28	S.R.
3	38	28	S.R.	7	2 13	28	S.R.
4	29 0 0 17	28	S.R.	8	52	28	S.R.

In the fourth column the letter O stands for Imperial Ordinary.
E ,, ,, Empress.
S ,, ,, Sovereign.
SR ,, ,, Ilford Special Rapid.

With the astrographic telescope 12 stars are shown on a number of plates, and seven stars on all but three (Nos. 13, 14 and 19). Of the eight plates taken with the 4-inch lens, seven show seven stars, but No. 6, which was taken through cloud, does not show any.

The following table of temperatures, communicated by Dr. MORIZE, and converted into the Fahrenheit scale, shows how slight the fall was during totality, probably owing to the large amount of cloud in the earlier stages which checked the usual daily rise.

G.M.T.	Ther.	G.M.T.	Ther.	G.M.T.	Ther.	G.M.T.	Ther.
d. h. m.	°	d. h. m.	°	d. h. m.	°	d. h. m.	°
28 22 45	82·4	28 23 30	80·6	29 0 15	82·0	29 1 0	83·8
23 0	84·2	45	82·4	30	82·4	15	84·2
15	82·4	29 0 0	80·6	45	83·1	30	84·2

15. On June 7, having completed the development, we left Sobral for Fortaleza, returning on July 9 for the purpose of securing comparison plates of the eclipse field.

Before our departure we dismounted the mirrors and driving clocks which were brought into the house to avoid the exposure to dust. The telescopes and cœlostats were left *in situ*. Before removing the mirrors we marked their positions in their cells so that they could be replaced in exactly the same position.

After our return to Sobral the mirrors and clocks were remounted; the photography of the eclipse field was commenced on the morning of July 11 (civil). The difficulty of finding the field with the cœlostats was overcome by making a rough hour circle on the heads of the cœlostats out of millimetre paper.

The following is the list of exposures made on the field for comparison with the eclipse photographs :—

Astrographic Telescope.

Ref. No.	Date.	G.M.T.	No. of exposures.	Duration.	Altitude.
		h. m.		s.	°
11_1	July 10	20 5	3	5	28·9
11_2		20 16	2	5	31·1
11_3		20 21	1	5	32·2
14_1	July 13	20 13	3	5	33·7
14_2		20 17	2	5	34·5
14_3		20 19	2	5	34·9
15_1	July 14	20 15	3	5	34·9
15_2		20 20	2	5	36·1
15_3		20 23	2	5	36·6
17_1	July 16	20 2	4	3	33·8
17_2		20 15	3	3	36·6
17_3		20 23	2	3	38·3
17_4		20 25	2	5	38·8
18_1	July 17	19 50	3	4	32·8
18_2		20 1	2	4	34·4
18_3		20 20	3	4	38·6
18_4		20 25	2	3	39·5

4-inch Telescope.

Ref. No.	Date.	G.M.T.	No. of exposures.	Duration.	Altitude.
		h. m.		s.	°
14_1	July 13	20 7	2	25	32·4
14_2		20 16	2	20	34·3
15_1	July 14	20 17	2	20	35·4
15_2		20 22	2	20	36·4
17_1	July 16	20 6	3	15	34·7
17_2		20 24	2	15	38·6
18_1*	July 17	19 57	3	20	33·6
18_2		20 24	2	20	39·2

The reference numbers follow the civil dates.

* The 4-inch plate, No. 18_1, was taken through the glass (see § 17, *infra*) to facilitate the measurement, and is referred to as the scale plate.

Thermometer readings, July 10, 74°·4; July 13, 73°·7; July 14, 71°·9; July 16, 72°·3; July 17, 72°·3.

By July 18 we had obtained a sufficient number of reference photographs. Dismantling of the instruments was commenced, and the packing was completed on July 21. We left Sobral on July 22, leaving the packing cases in the hands of Messrs. NICOLAU and CARNEIRO to be forwarded at the earliest opportunity, and arrived at Greenwich on August 25.

The observers wish to record their obligations to Mr. CHARLES BOOTH and the officers of the "Booth" Line for facilitating their journeys to and from their station at a difficult time.

PHOTOGRAPHS TAKEN WITH THE 4-INCH OBJECT GLASS.

16. These photographs were taken on 10 × 8-inch plates. By suitably mounting the camera it was made possible to obtain seven stars on the photographs, viz., Nos. 2, 3, 4, 5, 6, 10 and 11 of the table in § 5. Of the eight photographs taken during the eclipse seven gave measurable images of these stars, the other plate (No. 6) taken through cloud only showing a picture of the prominences.

Plates of the same field taken under nearly similar conditions as regards altitude were taken on July 14, 15, 17 and 18 (civil date). Of these photographs, the second taken on July 14 with two exposures (referred to as 14_{2a} and 14_{2b}), two photographs taken on July 15 (referred to as 15_1 and 15_2), two on July 17 (17_1 and 17_2), and the second photograph on July 18 (18_2) were measured for comparison with the eclipse plates.

17. The micrometer at the Royal Observatory is not suitable for the direct comparison of plates of this size. It was therefore decided to measure each plate by placing, film to film upon it, another photograph of the same region reversed by being taken through the glass. A photograph for this purpose was taken on July 18. This plate is regarded merely as an intermediary between the eclipse plates and comparison plates and is referred to as the scale plate, being used simply as a scale providing points of reference. In all cases measurement was made through the glass of the scale plate, adjusted on the eclipse or comparison plate which was being measured, so that the separation of the images on the two plates did not exceed one-third of a millimetre. The plates were held together by clips which ensured contact over the whole surface. This method of measurement was found to be very convenient. Each plate was measured in two positions, being reversed through 180 degrees, and the accordance of the result showed that the method of measurement was entirely satisfactory.

The measures, both direct and reversed, were made by two measurers (Mr. DAVIDSON and Mr. FURNER), and the means taken. There was no sensible difference between the measurers, which is satisfactory, as it affords evidence of the similarity of the images on the eclipse and comparison and scale plates.

The value of the micrometer screws (both in R.A. and Decl.) is 6″·25.

18. The results of the measures are as follows :—

TABLE II.—Eclipse Plates—Scale.

No. of Star.	I.		II.		III.		IV.		V.		VII.		VIII.	
	D*x*.	D*y*.	D*x*.	D*y*.	D*x*.	D*y*.	D*x*.	D*y*.	D*x*.	D*y*.	D*x*.	D*y*.	D*x*.	D*y*.
	r	*r*	*r*	*r*	*r*	*r*	*r*	*r*	*r*	*r*	*r*	*r*	*r*	*r*
11	—1·411	—0·554	—1·416	—1·324	+0·592	+0·956	+0·563	+1·238	+0·406	+0·970	—1·456	+0·964	—1·285	—1·195
5	—1·048	—0·338	—1·221	—1·312	+0·756	+0·843	+0·683	+1·226	+0·468	+0·861	—1·267	+0·777	—1·152	—1·332
4	—1·216	+0·114	—1·054	—0·944	+0·979	+1·172	+0·849	+1·524	+0·721	+1·167	—1·028	+1·142	—0·927	—0·930
3	—1·237	+0·150	—1·079	—0·862	+0·958	+1·244	+0·861	+1·587	+0·733	+1·234	—1·010	+1·185	—0·897	—0·894
6	—1·342	+0·124	—1·012	—0·932	+1·052	+1·197	+0·894	+1·564	+0·798	+1·130	—0·888	+1·125	—0·838	—0·937
10	—1·289	+0·205	—0·999	—0·948	+1·157	+1·211	+0·934	+1·522	+0·864	+1·119	—0·820	+1·072	—0·768	—0·964
2	—0·789	+0·109	—0·733	—1·019	+1·256	+0·924	+1·177	+1·373	+0·995	+0·935	—0·768	+0·892	—0·585	—1·166
	—1·500*	—0·554	—1·500	—1·324	+0·500	+0·843	+0·500	+1·226	+0·400	+0·861	—1·500	+0·777	—1·300	—1·322

COMPARISON Plates—Scale.

No. of Star.	14_{2a}.		14_{2b}.		15_1.		15_2.		17_1.		17_2.		18_2.	
	D*x*.	D*y*.	D*x*.	D*y*.	D*x*.	D*y*.	D*x*.	D*y*.	D*x*.	D*y*.	D*x*.	D*y*.	D*x*.	D*y*.
	r	*r*	*r*	*r*	*r*	*r*	*r*	*r*	*r*	*r*	*r*	*r*	*r*	*r*
11	—0·478	—0·109	+0·967	+1·170	+1·098	+1·228	+0·725	+0·830	—1·073	—1·330	+1·242	—0·302	—1·188	—1·572
5	—0·544	—0·204	+1·013	+1·192	+0·899	+1·232	+0·692	+0·938	—1·072	—1·075	+1·161	—0·224	—1·195	—1·432
4	—0·368	—0·136	+1·030	+1·249	+1·133	+1·086	+0·725	+0·854	—1·296	—1·031	+1·354	—0·281	—1·165	—1·454
3	—0·350	—0·073	+1·044	+1·305	+1·164	+1·114	+0·732	+0·893	—1·278	—1·014	+1·342	—0·261	—1·178	—1·394
6	—0·317	—0·144	+0·980	+1·319	+1·244	+1·012	+0·714	+0·824	—1·375	—1·052	+1·363	—0·390	—1·165	—1·473
10	—0·272	—0·146	+0·997	+1·327	+1·249	+0·960	+0·722	+0·831	—1·424	—1·038	+1·370	—0·423	—1·164	—1·476
2	—0·396	—0·182	+1·102	+1·289	+0·969	+1·052	+0·734	+0·941	—1·236	—0·909	+1·278	—0·328	—1·164	—1·335
	—0·552*	—0·206	+0·967	+1·170	+0·899	+0·960	+0·690	+0·824	—1·424	—1·330	+1·161	—0·423	—1·195	—1·572

* The numbers —1·500, —0·554, &c., given below the line, were taken out to make the values of D*x*, D*y* small and positive for arithmetical convenience.

19. The values of Dx and Dy were equated to expressions of the form

$$ax + by + c + \alpha E_x \ (= Dx)$$

and

$$dx + ey + f + \alpha E_y \ (= Dy),$$

where x, y are the co-ordinates of the stars given in Table I., and E_x, E_y are coefficients of the gravitational displacement.

The quantities c and f are corrections to zero, depending on the setting of the scale plate on the plate measured, a and e are differences of scale value, while b and d depend mainly on the orientation of the two plates. The quantity α denotes the deflection at unit distance (*i.e.*, 50′ from the sun's centre), so that αE_x and αE_y are the deflection in R.A. and Decl. respectively of a star whose co-ordinates are x and y.

The left-hand sides of the equation for the seven stars shown are :—

No.	Right Ascension.	Declination.
11	$c-0\cdot160b-1\cdot261a-0\cdot587\alpha$	$f-1\cdot261d-0\cdot160e+0\cdot036\alpha$
5	$c-1\cdot107b-0\cdot160a-0\cdot557\alpha$	$f-0\cdot160d-1\cdot107e-0\cdot789\alpha$
4	$c+0\cdot472b+0\cdot334a-0\cdot186\alpha$	$f+0\cdot334d+0\cdot472e+1\cdot336\alpha$
3	$c+0\cdot360b+0\cdot348a-0\cdot222\alpha$	$f+0\cdot348d+0\cdot360e+1\cdot574\alpha$
6	$c+1\cdot099b+0\cdot587a+0\cdot080\alpha$	$f+0\cdot587d+1\cdot099e+0\cdot726\alpha$
10	$c+1\cdot321b+0\cdot860a+0\cdot158\alpha$	$f+0\cdot860d+1\cdot321e+0\cdot589\alpha$
2	$c-0\cdot328b+1\cdot079a+1\cdot540\alpha$	$f+1\cdot079d-0\cdot328e-0\cdot156\alpha$

20. Normal equations formed from these equations of condition are as follows :—

TABLE III.—Eclipse Plates—Right Ascension.

					I.	II.	III.	IV.	V.	VII.	VIII.
+7·000c	+1·657b	+1·787a	+0·226α	=	+2·159	+2·986	+3·250	+2·461	+2·185	+3·263	+2·648
	+4·664	+2·089	+0·335	=	−0·063	+0·986	+1·320	+0·866	+1·051	+1·464	+1·130
		+4·094	+2·534	=	+1·034	+1·689	+1·866	+1·469	+1·480	+1·972	+1·723
			+3·142	=	+0·712	+0·919	+0·924	+0·860	+0·844	+0·930	+0·973
	+4·271b	+1·666a	+0·281α	=	−0·575	+0·278	+0·550	+0·283	+0·533	+0·691	+0·502
		+3·683	+2·476	=	+0·483	+0·928	+1·037	+0·841	+0·923	+1·140	+1·048
			+3·135	=	+0·643	+0·823	+0·820	+0·781	+0·774	+0·826	+0·888
		+2·988a	+2·366α	=	+0·707	+0·820	+0·822	+0·731	+0·715	+0·871	+0·852
			+3·116	=	+0·681	+0·805	+0·784	+0·762	+0·739	+0·780	+0·855
			+1·242α	=	+0·121	+0·156	+0·133	+0·183	+0·173	+0·090	+0·180
			α	=	+0·098	+0·126	+ ·107	+0·148	+0·140	+0·073	+0·145
			a	=	+0·158	+0·174	+0·189	+0·127	+0·128	+0·233	+0·169
			b	=	−0·203	−0·011	+0·048	+0·007	+0·042	+0·066	+0·042

Table IV.—Comparison Plates—Right Ascension.

					14_{2a}.	14_{2b}.	15_1.	15_2.	17_1.	17_2.	18_2.
+7·000*c*	+1·657*b*	+1·787*a*	+0·226α	=	+1·190	+0·364	+1·463	+0·214	+1·214	+0·983	+0·146
	+4·664	+2·089	+0·335	=	+0·700	+0·017	+0·992	+0·078	−0·340	+0·603	+0·083
		+4·094	+2·535	=	+0·638	+0·220	+0·499	+0·073	−0·172	+0·450	+0·085
			+3·142	=	+0·253	+0·159	−0·029	+0·037	−0·164	+0·105	+0·041
	+4·271*b*	+1·666*a*	+0·281α	=	+0·418	−0·069	+0·645	+0·027	−0·627	+0·370	+0·048
		+3·683	+2·476	=	+0·334	+0·127	+0·126	+0·018	−0·481	+0·199	+0·048
			+3·135	=	+0·215	+0·147	−0·076	+0·030	−0·203	+0·074	+0·036
		+2·988*a*	+2·366α	=	+0·172	+0·154	−0·126	+0·007	−0·236	+0·055	+0·029
			+3·116	=	+0·188	+0·152	−0·119	+0·028	−0·162	+0·050	+0·033
			+1·242α	=	+0·052	+0·030	−0·019	+0·022	+0·025	+0·006	+0·010
			α	=	+0·042	+0·024	−0·015	+0·018	+0·020	+0·005	+0·008
			a	=	+0·024	+0·032	−0·030	−0·012	−0·094	+0·014	+0·003
			b	=	+0·086	−0·030	+0·164	+0·012	−0·111	+0·081	+0·010

Table V.—Eclipse Plates—Declination.

+7·000*f*	+1·787*d*	+1·657*e*	+3·316α	=	+3·688	+1·927	+1·646	+1·452	+1·389	+1·718	+1·906
	+4·094	+2·089	+1·840	=	+2·200	+1·168	+0·719	+0·823	+0·555	+0·610	+0·840
		+4·664	+3·694	=	+1·860	+1·159	+1·129	+0·984	+0·874	+1·023	+1·193
			+5·784	=	+2·657	+1·681	+1·535	+1·361	+1·335	+1·545	+1·707
	+3·638*d*	+1·666*e*	+0·994α	=	+1·260	+0·677	+0·299	+0·453	+0·201	+0·172	+0·354
		+4·271	+2·908	=	+0·986	+0·702	+0·739	+0·640	+0·545	+0·616	+0·741
			+4·212	=	+0·909	+0·768	+0·755	+0·673	+0·677	+0·731	+0·804
		+3·508*e*	+2·453α	=	+0·409	+0·392	+0·602	+0·431	+0·453	+0·537	+0·579
			+3·941	=	+0·565	+0·583	+0·673	+0·549	+0·622	+0·684	+0·707
			+2·224α	=	+0·279	+0·309	+0·252	+0·247	+0·305	+0·308	+0·302
			α	=	+0·126	+0·139	+0·114	+0·111	+0·137	+0·139	+0·136
			e	=	+0·029	+0·015	+0·092	+0·045	+0·033	+0·056	+0·070
			d	=	+0·299	+0·141	+0·009	+0·074	+0·003	−0·016	+0·028

Table VI.—Comparison Plates—Declination.

+7·000*f*	+1·787*d*	+1·657*e*	+3·316α	=	+0·446	+0·661	+0·964	+0·343	+1·861	+0·752	+0·868
	+4·094	+2·089	+1·840	=	+0·060	+0·420	−0·156	+0·140	+1·038	+0·041	+0·476
		+4·664	+3·694	=	+0·202	+0·394	−0·203	−0·117	+0·526	−0·110	+0·122
			+5·784	=	+0·380	+0·482	+0·220	+0·044	+1·004	+0·296	+0·419
	+3·638*d*	+1·666*e*	+0·994α	=	−0·054	+0·251	−0·402	+0·053	+0·563	+0·151	+0·255
		+4·271	+2·908	=	+0·096	+0·237	−0·431	−0·198	+0·085	−0·288	−0·084
			+4·212	=	+0·168	+0·169	−0·237	−0·119	+0·122	−0·060	+0·008
		+3·508*e*	+2·453α	=	+0·121	+0·122	−0·247	−0·222	−0·173	−0·219	−0·201
			+3·941	=	+0·183	+0·100	−0·127	−0·133	−0·032	−0·019	−0·062
			+2·224α	=	+0·098	+0·015	+0·046	+0·022	+0·089	+0·134	+0·079
			α	=	+0·044	+0·007	+0·021	+0·010	+0·040	+0·060	+0·036
			e	=	+0·004	+0·030	−0·085	−0·070	−0·077	−0·104	−0·082
			d	=	−0·028	+0·054	−0·077	+0·044	+0·179	−0·010	+0·098

21. The values of α are collected in Table VII :—

TABLE VII.

Right Ascension.		Declination.	
Eclipse – Scale.	Comparison – Scale.	Eclipse – Scale.	Comparison – Scale.
r	r	r	r
+0·098	+0·042	+0·126	+0·044
+0·126	+0·024	+0·139	+0·007
+0·107	−0·015	+0·114	+0·021
+0·148	+0·018	+0·111	+0·010
+0·140	+0·020	+0·137	+0·040
+0·073	+0·005	+0·139	+0·060
+0·145	+0·008	+0·136	+0·036
Mean +0·120	+0·015	+0·129	+0·031

By subtracting the α of the comparison plates the scale plate is eliminated, and we derive from right ascensions $\alpha = +0^r\cdot105$ and from declinations $\alpha = +0^r\cdot098$.

Reference to the normal equations shows that the declination result is of double the weight of that from the right ascensions.

Thus

$$\alpha = +0^r\cdot100 = +0''\cdot625.$$

This is at a distance 50′ from the sun's centre. At the time of the eclipse the sun's radius was 15′·8 ; thus the deflection at the limb is 1″·98.

The range in the values of α is attributable to the errors inherent to the star images of the different plates, and cannot be reduced by further measurement. The mean values $+0^r\cdot015$ and $0^r\cdot031$ arise from the errors in the intermediary scale plate.

22. The probable error of the result judging from the accordance of the separate determinations is about 6 per cent. It is desirable to consider carefully the possibility of systematic error. The eclipse and comparison photographs were taken under precisely similar instrumental conditions, but there is the difference that the eclipse photographs were taken on the day of May 29, and the comparison photographs on nights between July 14 and July 18. A very satisfactory feature of the photographs is the essential similarity of the star images on the two sets of photographs.

The satisfactory accordance of the eclipse and comparison plates is shown by a study of the plate constants. The following corrections for differential refraction and aberration are calculated from the times and dates of exposure.

	a.	*c.*	*b.*	*d.*
	r	*r*	*r*	*r*
Eclipse plates	+0·240	+0·168	+0·062	+0·062
Scale plate	+0·423	+0·207	+0·096	+0·096
Comparison 14_{2a}	+0·409	+0·207	+0·091	+0·091
,, 14_{2b}	+0·409	+0·207	+0·091	+0·091
,, 15_1	+0·390	+0·207	+0·087	+0·087
,, 15_2	+0·370	+0·202	+0·087	+0·087
,, 17_1	+0·399	+0·207	+0·091	+0·091
,, 17_2	+0·337	+0·202	+0·077	+0·077
,, 18_2	+0·327	+0·202	+0·072	+0·072

When these are applied to the values of the constants found from the normal equations, we find the following values of the scale of the several photographs and their orientation relative to the scale plate :—

	Scale Value.		Orientation.		Adopted Scale Orientation.	
	From x.	From y.	From x.	From y.		
	r	*r*				*r*
Eclipse I.	−0·025	−0·010	−0·237	−0·265	0·000	−0·251
,, II.	−0·009	−0·024	−0·045	−0·107	0·000	−0·076
,, III.	+0·006	+0·053	+0·014	+0·025	0·000	+0·020
,, IV.	−0·056	+0·006	−0·027	−0·040	0·000	−0·034
,, V.	−0·055	−0·006	+0·008	+0·031	0·000	+0·020
,, VII.	+0·050	+0·017	+0·032	+0·050	0·000	+0·041
,, VIII.	−0·014	+0·031	+0·008	+0·006	0·000	+0·007
Comparison 14_{2a}	+0·010	+0·004	+0·081	+0·033	+0·013	+0·057
,, 14_{2b}	+0·008	+0·030	−0·035	−0·049	+0·013	−0·042
,, 15_1	−0·063	−0·085	+0·155	+0·086	−0·084	+0·120
,, 15_2	−0·065	−0·075	+0·003	−0·035	−0·084	−0·016
,, 17_1	−0·118	−0·077	−0·116	−0·174	−0·084	−0·145
,, 17_2	−0·072	−0·109	+0·062	+0·029	−0·084	+0·046
,, 18_2	−0·093	−0·087	−0·014	−0·074	−0·084	−0·044

The agreement in the scale values obtained from x and y is satisfactory. There appears to be a small difference in the orientations as derived from the two directions in the comparison plates. This is, however, of small importance in the determination of α. There is a difference of scale value from July 15–18 shown in both co-ordinates. For the purpose of exhibiting the gravitational displacements, residuals have been computed using adopted values for the scale and orientation given above, along with the calculated corrections for differential refraction and aberration. This has the advantage of reducing the number of constants employed in the reduction of the plates, and lessens the possibility of masking any discordances, though greater irregularities necessarily appear when four arbitrary constants instead of six are used in the reduction of each plate. The quantities are converted from revolutions to seconds of arc, as the more familiar unit facilitates judgment of the results.

TABLE VIII.—Comparison of the Eclipse and Comparison Photographs with the Scale Plate, after Correction for Differential Refraction and Aberration, Orientation and Change of Scale.

No. of Star.	I.	II.	III.	IV.	V.	VII.	VIII.	Mean.
	ECLIPSE Plates—Right Ascension.							
	″	″	″	″	″	″	″	″
11	−0·18	−0·51	−0·46	−0·07	−0·04	−0·72	−0·43	−0·34
5	−0·45	−0·81	−0·38	−0·58	−0·60	−0·36	−0·62	−0·54
4	+0·08	+0·11	−0·08	−0·11	−0·11	−0·16	−0·18	−0·06
3	−0·23	−0·11	−0·19	−0·05	−0·02	−0·02	−0·01	−0·09
6	−0·14	+0·23	−0·09	−0·11	−0·13	+0·13	−0·08	−0·03
10	+0·17	+0·06	+0·14	−0·18	−0·11	+0·14	−0·01	+0·03
2	+0·75	+1·03	+1·06	+1·09	+1·01	+0·98	+1·30	+1·03
	ECLIPSE PLATES—Declination.							
	″	″	″	″	″	″	″	″
11	0·00	−0·08	−0·03	+0·02	+0·17	+0·16	+0·01	+0·03
5	−0·38	−0·54	−0·61	−0·30	−0·39	−0·73	−0·81	−0·54
4	+1·19	+1·04	+1·03	+0·98	+1·11	+1·19	+1·24	+1·11
3	+1·42	+1·58	+1·50	+1·39	+1·55	+1·49	+1·49	+1·49
6	+0·65	+0·79	+1·01	+0·97	+0·71	+0·95	+1·01	+0·87
10	+0·62	+0·46	+1·03	+0·54	+0·56	+0·58	+0·74	+0·65
2	+0·01	+0·25	−0·40	−0·09	−0·22	−0·14	−0·17	−0·11

	14_{2a}.	14_{2b}.	15_1.	15_2.	17_1.	17_2.	18_2.	Mean.
	COMPARISON Plates—Right Ascension.							
	″	″	″	″	″	″	″	″
11	−0·19	−0·24	−0·23	−0·28	+0·11	−0·19	−0·02	−0·15
5	−0·42	+0·16	−0·36	−0·32	−0·24	−0·33	−0·26	−0·25
4	−0·01	+0·03	−0·01	+0·05	−0·04	+0·23	+0·08	+0·05
3	+0·14	+0·09	+0·28	+0·10	−0·03	+0·21	−0·01	+0·11
6	+0·02	−0·18	+0·26	+0·06	+0·13	+0·03	+0·14	+0·07
10	+0·17	−0·06	+0·20	+0·18	+0·13	−0·02	+0·15	+0·11
2	+0·31	+0·18	−0·16	+0·22	−0·04	+0·08	−0·06	+0·08
	COMPARISON Plates—Declination.							
	″	″	″	″	″	″	″	″
11	−0·07	+0·08	−0·26	−0·04	−0·26	−0·18	−0·16	−0·13
5	−0·23	−0·03	+0·03	0·00	−0·19	+0·03	−0·20	−0·08
4	+0·23	+0·05	+0·29	+0·18	+0·45	+0·53	+0·23	+0·28
3	+0·64	+0·41	+0·42	+0·36	+0·48	+0·60	+0·54	+0·49
6	+0·22	+0·36	+0·33	+0·26	+0·41	+0·21	+0·32	+0·30
10	+0·28	+0·32	+0·31	+0·36	+0·36	+0·15	+0·29	+0·30
2	+0·25	+0·14	+0·18	+0·21	+0·09	−0·03	+0·27	+0·16

Subtracting the results of the comparison plates, so as to eliminate the errors arising from the intermediary scale plate we find for the displacements of the different stars, as compared with those as given by EINSTEIN'S Theory, with value 1″·75 at the sun's limb :—

No. of Star.	Displacement in Right Ascension.		Displacement in Declination.	
	Observed.	Calculated	Observed.	Calculated.
	″	″	″	″
11	−0·19	−0·32	+0·16	+0·02
5	−0·29	−0·31	−0·46	−0·43
4	−0·11	−0·10	+0·83	+0·74
3	−0·20	−0·12	+1·00	+0·87
6	+0·10	+0·04	+0·57	+0·40
10	−0·08	+0·09	+0·35	+0·32
2	+0·95	+0·85	−0·27	−0·09

PHOTOGRAPHS TAKEN WITH THE ASTROGRAPHIC OBJECT GLASS.

23. As stated above these photographs were taken with the astrographic object glass stopped down to 8 inches, mounted in a steel tube and fed by a 16-inch cœlostat. From many years' experience with the object glass at Greenwich it is certain that, when the object glass is mounted in a steel tube, the change of scale over a range of temperature of 10° F. should be insignificant, and the definition should be very good. It was realised that this high standard would not be obtained with the glass used in conjunction with the cœlostat taken to Brazil, but nevertheless the results shown when the plates were developed were very disappointing. The images were diffused and apparently out of focus, although on the night of May 27 the focus was good.* Worse still, this change was temporary, for without any change in the adjustments, the instrument had returned to focus when the comparison plates were taken in July.

These changes must be attributed to the effect of the sun's heat on the mirror, but it is difficult to say whether this caused a real change of scale in the resulting photographs or merely blurred the images.

The photographs were measured in the astrographic duplex micrometer, the eclipse photographs being directly compared with the comparison plates taken in July. All

* The following note made at the time is quoted in full :—" May 30, 3 a.m., four of the astrographic plates were developed, and when dry examined. It was found that there had been a serious change of focus, so that, while the stars were shown, the definition was spoilt. This change of focus can only be attributed to the unequal expansion of the mirror through the sun's heat. The readings of the focussing scale were checked next day, but were found unaltered at 11·0 mm. It seems doubtful whether much can be got from these plates."

the stars shown were measured. They were reduced by the same method as that employed for the "4-inch" photographs. With the exception of plates Nos. 15 and 16, taken through clouds, the stars numbered 3, 4, 5, 6, 10, 11 and 12 are shown on all the plates; the fainter stars 2, 7, 8 and 9 are sometimes shown, but No. 1, which is very near the sun, is always drowned in the corona. These plates were only measured in declination, as the right ascensions were of little weight.

24. In the following table is given the value of α, the constant of the gravitational displacement, as calculated from the measures; the apparent difference of scale e between the eclipse and comparison plates; d the difference of orientation of the plates given by the measures of y, and depending on the adjustment of the plates in the measuring machine.

TABLE IX.

($1^r = 12''\cdot 3$).

No. of Eclipse Plate.	Ref. No. of Comparison Plate.	No. of Stars.	Values of d, e, α in Revolutions at 50′ Distance.			α at Sun's Limb in Arc.
			d.	e.	α.	
			r	r	r	″
1	18_4	7	+0·051	+0·089	+0·033	+1·28
2	18_4	11	−0·009	+0·059	+0·025	+0·97
3	18_4	8	−0·074	+0·101	+0·028	+1·09
4	18_4	11	−0·168	+0·091	+0·033	+1·28
5	11_3	10	+0·094	+0·076	+0·025	+0·97
6	11_3	11	+0·186	+0·082	+0·021	+0·82
{ 7	14_3	12	+0·006	+0·119	0·000	0·00
{ 7	18_3	7	−0·054	+0·166	0·000	0·00
8	14_3	10	+0·093	+0·064	+0·021	+0·82
9	17_4	7	−0·096	+0·129	+0·008	+0·31
10	17_4	10	+0·090	+0·045	+0·026	+1·01
11	11_1	10	+0·073	+0·061	+0·032	+1·24
{ 12	11_1	11	−0·009	+0·102	+0·049	+1·91
{ 12	17_2	7	−0·102	+0·114	+0·019	+0·74
15	15_3	6	+0·111	+0·036	+0·018	+0·70
16	15_3	7	−0·002	+0·037	+0·018	+0·70
17	17_2	8	−0·022	+0·109	+0·012	+0·47
18	17_2	7	+0·045	0·000	+0·030	+1·17
		Mean		+0·082	+0·022	+0·86

Thus the mean value of α obtained from all the astrographic plates is $0''\cdot 86$, a figure considerably less than that obtained from the 4-inch photographs.

25. Reference to the diagram shows that the measurement of displacement depends essentially on the position of the stars Nos. 3 and 4 relative to 5 on one side and 6 and 10 on the other. These are all bright stars, and in this respect their images are

more comparable than are the images of the fainter stars. The measures of these stars are given in the following table :—

No. of Eclipse Plate.	Measured Values of Dy for Stars Nos.— 5	4	3	6	10	No. of Eclipse Plate.	Measured Values of Dy for Stars Nos.— 5	4	3	6	10
	r	r	r	r	r		r	r	r	r	r
1	−0·051	+0·175	+0·169	+0·201	+0·235	9	−0·059	+0·121	+0·109	+0·205	+0·180
2	+0·558	+0·656	+0·724	+0·668	+0·702	10	+0·033	+0·270	+0·188	+0·258	+0·280
3	+0·124	+0·285	+0·286	+0·274	+0·355	11	+0·025	+0·215	+0·210	+0·233	+0·274
4	+0·111	+0·222	+0·247	+0·231	+0·167	12	−0·068	+0·144	+0·124	+0·160	+0·167
5	+0·034	+0·228	+0·232	+0·218	+0·308	15	−0·038	+0·138	+0·107	+0·172	—
6	+0·164	+0·488	+0·478	+0·557	+0·637	16	−0·050	+0·076	+0·046	+0·127	+0·073
7	−0·051	+0·156	+0·162	+0·250	+0·279	17	−0·071	+0·104	+0·081	+0·186	+0·164
8	+0·108	+0·330	+0·314	+0·376	+0·397	18	+0·016	+0·092	+0·109	+0·099	+0·084

The equations given by these stars are

$$-0\cdot160d - 1\cdot107e - 0\cdot789a + f = Dy_5 \quad (1)$$
$$+0\cdot334d + 0\cdot472e + 1\cdot336a + f = Dy_4 \quad (2)$$
$$+0\cdot348d + 0\cdot360e + 1\cdot574a + f = Dy_3 \quad (3)$$
$$+0\cdot587d + 1\cdot099e + 0\cdot726a + f = Dy_6 \quad (4)$$
$$+0\cdot860d + 1\cdot321e + 0\cdot589a + f = Dy_{10} \quad (5)$$

The mean of (4) and (5) added to (1) gives

$$+0\cdot564d + 0\cdot103e - 0\cdot131a + 2f = Dy_5 + \tfrac{1}{2}(Dy_6 + Dy_{10}).$$

While the sum of (2) and (3) gives.

$$+0\cdot682d + 0\cdot832e + 2\cdot910a + 2f = Dy_3 + Dy_4.$$

Subtracting these we get

$$3\cdot041a + 0\cdot729e + 0\cdot118d = Dy_3 + Dy_4 - Dy_5 - \tfrac{1}{2}(Dy_6 + Dy_{10}).$$

This equation has a small coefficient for e and a very small one for d.

Calculating the quantities on the right-hand side, assuming e to be the same for all the plates, and substituting the values of d from the previous table, we find :—

	r			r	
$a + 0\cdot240e =$	$+0\cdot056$	1	$a + 0\cdot240e =$	$+0\cdot035$	9
$a + 0\cdot240e =$	$+0\cdot049$	2	$a + 0\cdot240e =$	$+0\cdot048$	10
$a + 0\cdot240e =$	$+0\cdot047$	3	$a + 0\cdot240e =$	$+0\cdot045$	11
$a + 0\cdot240e =$	$+0\cdot059$	4	$a + 0\cdot240e =$	$+0\cdot059$	12
$a + 0\cdot240e =$	$+0\cdot050$	5	$a + 0\cdot283e =$	$+0\cdot026$	15
$a + 0\cdot240e =$	$+0\cdot059$	6	$a + 0\cdot240e =$	$+0\cdot024$	16
$a + 0\cdot240e =$	$+0\cdot036$	7	$a + 0\cdot240e =$	$+0\cdot028$	17
$a + 0\cdot240e =$	$+0\cdot046$	8	$a + 0\cdot240e =$	$+0\cdot029$	18

In photograph No. 15, star 10 is not shown, and the equation is slightly modified. It may also be noticed that the values are somewhat smaller for Nos. 15 to 18.

The means of the 16 photographs treated in this manner give

$$a + 243e = + 0^r\cdot0435,$$

or with the value of the scale $0^r\cdot082$ from the previous table

$$a = + 0^r\cdot024 = 0''\cdot93 \text{ at the limb.}$$

It may be noticed that the change of scale arising from differences of refraction and aberration is $0^r\cdot020$. If this value of e be taken instead of $0^r\cdot082$ we obtain

$$a = + 0^r\cdot039 = + 1''\cdot52 \text{ at the sun's limb.}$$

The equations on p. 311 were also solved by least squares for each plate. There is a considerable range in the deduced values of a, as is to be expected when a and e are determined independently for each plate. The mean result for a is $0''\cdot99$, or very nearly the same as that already found.

The photographs taken with the astrographic telescope support those obtained by the "4-inch" to the extent that they show considerable outward deflection, but for the reasons already given are of much less weight.

IV. The Expedition to Principe.

(*Observers, Prof.* A. S. Eddington *and Mr.* E. T. Cottingham.)

26. The expedition left Liverpool on the "Anselm" on March 8, and travelled in company with the Sobral expedition as far as Madeira. It was necessary to wait there until April 9, when the journey was continued on the "Portugal," belonging to the Companhia Nacional de Navegação. The expedition landed at the small port of S. Antonio in the Isle of Principe on April 23.

Vice-Admiral Campos Rodrigues and Dr. F. Oom of the National Observatory, Lisbon, had kindly given us introductions, and everything possible was done by those on the island for the success of the work and the comfort of the observers. We were met on board by the Acting Administrator Sr. Vasconcélos, Sr. Carneiro, President of the Association of Planters, and Sr. Grageira, representing the Sociedade d'Agricultura Colonial, who made all necessary arrangements. The Portuguese Government dispensed with any customs examination of the baggage.

27. Principe is a small island belonging to Portugal, situated just north of the equator in the Gulf of Guinea, about 120 miles from the African coast. The extreme length and breadth are about 10 miles and 6 miles. Near the centre mountains rise to a height of 2500 feet, which generally attract heavy masses of cloud. Except for a certain amount of virgin forest, the island is covered with cocoa plantations. The

climate is very moist, but not unhealthy. The vegetation is luxuriant, and the scenery is extremely beautiful. We arrived near the end of the rainy season, but the *gravana*, a dry wind, set in about May 10, and from then onwards no rain fell except on the morning of the eclipse.

We were advised that the prospects of clear sky at the end of May were not very good, but that the best chance was on the north and west of the island. After inspecting two other sites on the property of the Sociedade d'Agricultura Colonial, we fixed on Roça Sundy, the headquarters of Sr. CARNEIRO's chief plantation. We were Sr. CARNEIRO's guests during our whole visit, and used freely his ample resources of labour and material at Sundy. We learnt later that he had postponed a visit to Europe in order to entertain us. We were also greatly indebted to his manager at Sundy, Sr. ATALAYA, with whom we lived for five weeks; his help and attention were invaluable. Mr. WRIGHT and Mr. LEWIS of the Cable Station kindly assisted us as interpreters when necessary.

Sundy is situated in the north-west of the island overlooking the sea at a height of 500 feet, and as far as possible from the cloud-gathering peaks. Our telescope was erected in a small walled enclosure adjoining the house, from which the ground sloped steeply down to the sea in the direction of the sun at eclipse. On the other side it was sheltered by a building. The approximate position was latitude 1° 40′ N., longitude 29m. 32s. E.

28. The baggage was brought to Sundy on April 28 mainly by tram, but with a break of about a kilometre, where it had to be transported through the wood by native carriers. After a week spent on the preparations, we returned to S. Antonio for the week, May 6–13, as it was undesirable to unpack the mirror so early in the damp climate. On our return to Sundy the installation and adjustments were soon completed, and the first check plates were taken on May 16. Meanwhile the gravana had begun, which, although there is no rain, is generally accompanied by increased cloud. There were, however, some days of clear sky, and the nights were usually clear.

The cœlostat was mounted on a stone pier built for the purpose. The clock weight fell into a pit below the clock deep enough to allow a run of 36 minutes without rewinding. Care was taken to use a particular part of the cœlostat-sector, considered to be the most perfect, in photographing the eclipse and the check field. The telescope (Oxford astrographic object-glass, see p. 295) rested on wooden V's near the two ends, the V's being supported on packing-cases; the one at the breech-end could be moved laterally to allow of different declination settings, and was marked with an approximate declination scale. A series of exposures of one second was made on a bright star to test whether there was any shake of the telescope after inserting the plate: no shake was detected even when the exposure was made immediately; but as a safeguard for the eclipse photographs a full second was allowed to elapse before beginning the exposure. The exposure was made by moving a cardboard screen unconnected with the instrument. The telescope pointed slightly downwards, and the tube was turned so as to give the right orientation to the plate, the lines of declination being two or three degrees inclined to the horizontal. A canvas screen was arranged to protect the tube and object-glass from the direct radiation of the sun.

The adjustments call for little comment. In view of the purpose of the observations, it was desirable to adjust the tilt of the object-glass and plate with special care. It was also important that the setting on the field should be nearly exact. The sun appeared on the eclipse day in sufficient time to allow of the setting being made by means of the solar image; but arrangements had been tested by which the correct field would have been obtained if it had been cloudy up to totality.* The telescope was focussed by trial photographs of stars, and owing to the uniform temperature of the island the focus was unchanged for day observations.

The object-glass was stopped down to 8 inches for the eclipse photographs and for all check and comparison photographs used in the reductions.

29. The days preceding the eclipse were very cloudy. On the morning of May 29 there was a very heavy thunderstorm from about 10 a.m. to 11.30 a.m.—a remarkable occurrence at that time of year. The sun then appeared for a few minutes, but the clouds gathered again. About half-an-hour before totality the crescent sun was glimpsed occasionally, and by 1.55 it could be seen continuously through drifting cloud. The calculated time of totality was from 2h. 13m. 5s. to 2h. 18m. 7s. G.M.T. Exposures were made according to the prepared programme, and 16 plates were obtained. Mr. COTTINGHAM gave the exposures and attended to the driving mechanism, and Prof. EDDINGTON changed the dark slides. It appears from the results that the cloud must have thinned considerably during the last third of totality, and some star images were shown on the later plates. The cloudier plates give very fine photographs of a remarkable prominence which was on the limb of the sun.

A few minutes after totality the sun was in a perfectly clear sky, but the clearance did not last long. It seems likely that the break-up of the clouds was due to the eclipse itself, as it was noticed that the sky usually cleared at sunset.

It had been intended to complete all the measurements of the photographs on the spot; but owing to a strike of the steamship company it was necessary to return by the first boat, if we were not to be marooned on the island for several months. By the intervention of the Administrator berths, commandeered by the Portuguese Government, were secured for us on the crowded steamer. We left Principe on June 12, and after transhipping at Lisbon, reached Liverpool on July 14.

30. The following is a list of the photographs, including the comparison photographs kindly taken for us by Mr. F. A. BELLAMY at Oxford, before the instrument was dismounted. All the eclipse photographs are given, though only W and X furnished

* The method depended on setting the cross-wires of the theodolite (attached to the cœlostat) on a terrestrial mark, and then starting the clock at a particular instant.

results. Of the other series, only the exposures actually used in the reductions are given.

List of Plates.

Check Field (R.A. 14h. 12m. 47s., Declination +20° 30′)

Ref.	Place.	Date.	Loc. Sid. T.	Exp.	Approx. Z.D.	Bar.	Ther.	Plate.
		1919.	h. m. s.	s.	°	m.	°	
a_1	Oxford	January 16	12 55 10	60	35	29·64	37·0	S.
b_1	,,	January 17	13 10 40	60	34	29·83	35·3	S.
c_1	,,	,,	13 54 55	60	31	29·83	35·3	S.
d_1	,,	,,	14 9 25	60	31	29·83	35·3	S.
e_1	,,	January 23	13 13 30	60	33	30·45	29·0	S.
			G.M.T.					
q_1	Principe	May 22	12 25 40	40	43	29·45	76·5	S.R.
r_1	,,	,,	12 31 20	40	45	29·45	76·5	S.R.
s_2	,,	,,	12 37 50	80	46	29·45	76·5	S.R.
v_1	,,	May 25	12 22 20	40	45	29·45	76·5	S.S.
w_1	,,	,,	12 26 20	40	46	29·45	76·5	S.S.

Notes.

Column 1.—The letter is marked on the original plates (preserved at Cambridge Observatory). The number refers to the exposure, disregarding exposures taken without the 8-inch stop.

Column 2.—The co-ordinates of Oxford Observatory are 5m. 3s. W., 51° 46′ N., and of the site at Principe, 29m. 32s. E., 1° 40′ N.

Column 4.—The mid-instant of the exposure is given. Times for check plates at Principe were only noted roughly. Times for the eclipse plates are deduced from the calculated time of totality, the interval from the end of one exposure to the beginning of the next being assumed uniform.

Column 7.—Readings at Principe were taken with an aneroid recording instrument, and therefore automatically reduced to the latitude of England. The barometer during our visit was practically constant except for a regular semi-diurnal wave of amplitude about 0·05 in.

Column 9.—Brand of Plate : S. = Imperial Sovereign, S.S. = Imperial Special Sensitive, S.R. = Ilford Special Rapid, E. = Ilford Empress. Backed plates were used at Principe.

Eclipse Field (R.A. 4h. 19m. 30s., Declination +21° 43′)

Ref.	Place.	Date.	Loc. Sid. T.	Exp.	Approx. Z.D.	Bar.	Ther.	Plate.
		1919.	h. m. s.	s.	°	m.	°	
D_1	Oxford	January 16	3 58 1	5	30	29·65	39·0	S.
G_1	,,	January 22	4 4 39	5	30	30·30	31·0	S.
H_1	,,	,,	4 34 28	5	30	30·30	31·0	S.
I_2	,,	,,	4 48 46	10	31	30·30	31·0	S.
K_2	,,	February 9	4 45 24	10	30	30·48	24·5	S.
			G.M.T.					
K	Principe	May 29	2 13 9	5	46	29·45	77·0	S.R.
L	,,	,,	2 13 28	10	46	29·45	77·0	S.R.
M	,,	,,	2 13 46	3	46	29·45	77·0	S.R.
N	,,	,,	2 14 1	5	46	29·45	77·0	E.
O	,,	,,	2 14 20	10	46	29·45	77·0	S.S.
P	,,	,,	2 14 44	15	46	29·45	77·0	S.S.
Q	,,	,,	2 15 6	5	46	29·45	77·0	S.R.
R	,,	,,	2 15 30	20	46	29·45	77·0	S.R.
S	,,	,,	2 15 53	3	46	29·45	77·0	S.S.
T	,,	,,	2 16 13	15	46	29·45	77·0	E.
U	,,	,,	2 16 37	10	46	29·45	77·0	S.R.
V	,,	,,	2 16 56	5	46	29·45	77·0	S.S.
W	,,	,,	2 17 15	10	46	29·45	77·0	S.
X	,,	,,	2 17 33	3	46	29·45	77·0	S.R.
Y	,,	,,	2 17 47	2	46	29·45	77·0	S.R.
Z	,,	,,	2 18 1	2	46	29·45	77·0	S.R.

Notes.

Columns 1 to 9. See previous page.

The large proportion of Ilford Special Rapid plates used at the eclipse was due to the fact that experience in developing the check plates showed that these suffered less than the others from the high temperature of the water (78° F.). Ice was generally available for the check plates through the kindness of Sr. Grageira; but the supply failed after the eclipse, and formalin was used to harden the films. This was unsatisfactory except for the I.S.R. plates, and so plates P, S, T, W were brought home undeveloped. The developing at Principe was done at night, and the drying was accelerated by use of alcohol.

The use of an 8-inch stop in front of the object-glass was suggested to us by Mr. Davidson, who showed that a great improvement of the images resulted; it was originally intended, however, to use the full aperture for part of totality. Early measures of check plates made at Principe soon convinced us that the results from the full aperture were greatly inferior, and we decided to rely entirely on the 8-inch aperture.

The Check Plates.

31. In addition to the eclipse field, a check field was photographed both at Oxford and at Principe. The field chosen included Arcturus, so that it was easily found with the cœlostat. Its declination was nearly the same as that of the eclipse field, and it was photographed at the same altitude at Principe in order that any systematic error, due to imperfections of the cœlostat mirror or other causes, might affect both sets of plates equally. The primary purpose was thus to check the possibility of systematic error arising from the different conditions of observation at Oxford and Principe, and from possible changes in the object-glass during transit. Unlike the Sobral expedition, we were not able to take comparison photographs of the eclipse field at Principe, because for us the eclipse occurred in the afternoon, and it would be many months before the field could be photographed in the same position in the sky before dawn. The check plates were therefore specially important for us.

As events turned out the check plates were important for another purpose, viz., to determine the difference of scale at Oxford and Principe. As shown in the report of the Sobral expedition, it is not necessary to know the scale of the eclipse photographs, since the reductions can be arranged so as to eliminate the unknown scale. If, however, a trustworthy scale is known and used in the reductions, the equations for the deflection have considerably greater weight, and the result depends on the measurement of a larger displacement. On surveying the meagre material which the clouds permitted us to obtain, it was evident that we must adopt the latter course; and accordingly the first step was to obtain from the check plates a determination of the scale of the Principe photographs.

32. All the measures were made by Prof. EDDINGTON with the Cambridge measuring machine.* An Oxford and a Principe plate were placed film to film so that the images of corresponding stars nearly coincided—this was possible because the Oxford plates were taken direct, and the Principe plates by reflection in the cœlostat mirror.

The small differences Δx and Δy, in the sense Principe − Oxford, were then measured for each star. Eight settings were made on each image; for half of them the field was rotated through 180 degrees by the reversion prism. Five pairs of plates were measured, and the measures are given in Table XI.

* 'Monthly Notices, R.A.S.,' vol. LXI, p. 444.

TABLE XI.—Check Plates, Measures.

Star.	Approx. Co-ords.		q_1-a_1.		w_1-b_1.		s_2-c_1.		r_1-d_1.		v_1-e_1.	
	x.	y.	Δx.	Δy.	Δx.	Δy.	Δx.	Δy.	Δx.	Δy.	Δx.	Δy.
1	1·41	20·31	4346	7180	3199	4259	6012	7375	3921	8796	5435	4399
2	5·89	12·74	3865	6405	3394	4129	4922	6132	3039	7440	5978	4170
4	9·46	11·13	3640	5932	3408	4118	4369	5366	2638	6776	5966	4441
5	12·00	6·84	3311	5590	—	—	3831	4752	1938	6156	—	4314
6	12·80	27·33	5415	6561	3192	5140	7689	5925	5379	7580	5032	5794
7	13·75	13·78	4076	5630	3496	4290	4891	4805	3101	6461	5906	4826
8	15·50	24·38	5125	6300	—	—	—	—	—	—	5139	5412
10	20·13	10·49	3965	4940	3679	4505	4656	3568	2866	5370	6398	5229
11	20·81	0·93	2874	4352	3876	3759	2845	2815	1238	4758	7268	4482
12	22·91	6·23	3685	4436	3931	4158	4039	2738	2270	4551	6765	5076
13	26·46	8·96	4222	4288	4045	4326	4724	2232	2720	4120	6836	5561

The unit for x and y is 5 millimetres, which is approximately equal to $5'$. The differences Δx, Δy are given in units of the fifth place of decimals $= 0''\cdot003$. The centre of the plate is near $x = 14$, $y = 14$.

Plate-constants were then calculated in the usual way, by the formulæ

$$\Delta x = ax + by + c$$

$$\Delta y = dx + ey + f$$

These were applied, and the residuals $\Delta_1 x$, $\Delta_1 y$ converted into arc are as follows:—

TABLE XII.—Check Plates, Residuals.

Star.	q_1-a_1.		w_1-b_1.		s_2-c_1.		r_1-d_1.		v_1-e_1.		Mean.	
	$\Delta_1 x$.	$\Delta_1 y$.	$\Delta_1 x$.	$\Delta_1 y$.	$\Delta_1 x$.	$\Delta_1 y$.	$\Delta_1 x$.	$\Delta_1 y$.	$\Delta_1 x$.	$\Delta_1 y$.	$\Delta_1 x$.	$\Delta_1 y$.
	$''$	$''$	$''$	$''$	$''$	$''$	$''$	$''$	$''$	$''$	$''$	$''$
1	−0·02	−0·02	+0·29	−0·34	+0·02	−0·07	−0·03	+0·22	+0·49	+0·01	+0·15	−0·04
2	+0·39	+0·15	+0·16	+0·14	+0·69	0·00	+0·69	−0·29	+0·10	−0·23	+0·41	−0·05
4	−0·14	−0·04	−0·16	+0·09	−0·38	−0·12	−0·02	−0·37	−0·54	+0·12	−0·25	−0·06
5	−0·08	+0·35	—	—	+0·25	+0·19	−0·21	−0·21	—	−0·01	−0·01	+0·08
6	−0·06	−0·10	−0·28	+0·27	−0·09	+0·14	−0·10	+0·12	+0·15	+0·49	−0·08	+0·18
7	−0·06	−0·28	−0·10	−0·16	−0·74	−0·09	−0·31	+0·02	−0·39	−0·12	−0·32	−0·13
8	−0·30	+0·34	—	—	—	—	—	—	−0·38	−0·68	−0·34	−0·17
10	−0·02	−0·10	−0·21	+0·52	−0·15	+0·16	+0·08	+0·25	−0·08	+0·34	−0·08	+0·23
11	−0·46	−0·01	−0·13	−0·22	−0·13	+0·11	−0·13	+0·71	+0·30	−0·28	−0·11	+0·06
12	+0·16	−0·14	+0·13	−0·04	+0·19	−0·06	+0·17	−0·09	−0·13	+0·08	+0·10	−0·05
13	+0·59	−0·12	+0·32	−0·26	+0·34	−0·25	−0·13	−0·38	+0·48	+0·28	+0·32	−0·15

The mean residual without regard to sign is $\pm 0''\cdot21$, from which the probable error of a determination of Δx or Δy is $\pm 0''\cdot22$.

Star 7 is much the brightest. Stars 1, 6, 11, 13 are rather bright. Stars 2, 4, 10, 12 are fainter and more comfortable to measure. Stars 5 and 8 are very faint. Arcturus is on the plates but is much too bright to measure. No measures have been rejected.

The determination of the deflection on the eclipse plates is based on the declinations (y), and the last column of Table XII. shows that on the check plates the y-comparisons are free from any serious systematic error.

Star 7 is of particular interest ; its position near the centre of the field corresponds to that of κ_1, κ_2 Tauri in the eclipse field, from which the greatest deflection is expected. The images (which are not quite round) have the same characteristic shape. Further, the brightness of No. 7 corresponds with, but exaggerates, the brightness of κ_1 Tauri which is the brightest star in the eclipse field. It is therefore a valuable check to find that its systematic error in declination is insignificant compared with the displacement (of the order of 1″) afterwards found for κ_1 and κ_2 Tauri.

The systematic errors in right ascension are larger (probably through imperfect driving of the clock). They may affect the displacement indirectly through the orientation constant, but with much reduced effect. Allowing for this reduction in importance there appears to be nothing to trouble about.

The primary purpose of the check plates is thus fulfilled. They show that photographs of a check field of stars taken at Oxford and Principe show none of the displacements which are exhibited by the photographs of the eclipse field taken under precisely similar instrumental conditions. The inference is that the displacements in the latter case can only be attributed to presence of the eclipsed sun in the field.

33. We turn now to the differences of scale between Oxford and Principe, which are given by the plate-constants a, b, d, e determined from the measures. As determined, these include the effects of differential refraction and aberration. The latter corrections were calculated for each plate by the usual formulæ and applied, so as to determine the corrected plate-constants, a', b', d', e' free from differential refraction and aberration. Due allowance was made for the change in the coefficient of refraction owing to the difference of barometer and temperature (about 40°) between Oxford and Principe. The results are as follows (in units of the fifth place of decimals) :—

TABLE XIII.—Check Plates, Plate-Constants.

Comparison.	Uncorrected.				Corrected.				
	a.	b.	d.	e.	a'.	b'.	d'.	e'.	$b'+d'$.
$q_1 - a_1$	+32·7	+101·0	− 87·8	+58·2	+32·7	+ 98·4	− 90·4	+32·1	+ 8·0
$w_1 - b_1$	+26·2	− 16·0	+ 25·9	+53·6	+30·4	− 22·5	+ 19·4	+31·4	− 3·1
$s_2 - c_1$	+31·5	+192·5	−173·5	+64·8	+35·8	+182·6	−183·4	+42·1	− 0·8
$r_1 - d_1$	+28·2	+165·0	−146·8	+69·8	+32·1	+157·8	−154·0	+45·0	+ 3·8
$v_1 - e_1$	+21·6	− 76·2	+ 70·6	+61·4	+25·2	− 80·5	+ 66·3	+35·7	−14·2
Mean					+31·2	—	—	+37·3	− 1·3

The sign of the results shows that the scale of the photographs is larger at Principe than at Oxford ; in fact the focus must have been set about 1·2 mm. further out (apart from any change of length compensated by expansion of the photographic plates). As the error in focussing was probably not more than 0·5 mm., the greater part of this shift must be due to the focal length of the lens combination increasing with temperature more rapidly than the linear expansion of the glass.

If the only difference were a change of focal length, we should have $a' = e'$. There is a fairly strong indication that e' is greater than a'. This is no doubt due to a change in the definition caused by the cœlostat mirror or by a shift of the object-glass lenses on the journey ; and, as it will presumably affect the eclipse plates in the same way, it is best to adopt the values of a' and e' as determined, rather than to take a mean. In so doing we shall at any rate not exaggerate the displacement, which depends mainly on the y-measures and is reduced by adopting too large a value of e'.*

The difference $b' - d'$ merely gives the relative orientation of the two plates as placed face to face. The sum $b' + d'$ practically vanishes, as it should do. However, for consistency we adopt the small value found.

From the internal discordances of our determination of e' (the most important of these constants) the probable error of the mean is $\pm 2 \cdot 1$. This, as shown later, will cause a probable error of our final determination of the deflection, reduced to the limb of the sun, of amount $\pm 0'' \cdot 14$, affecting all determinations systematically. Errors in the other constants have much smaller influence.

The Eclipse Plates.

34. The eclipse plates from K to S show no star images. After that the cloud lightened somewhat, and some images appear on the remaining plates. The sky was never clear and nothing fainter than $5' \cdot 5$ is shown. The cloud was variable in different parts of the plate, so that the brightness of the images varies erratically and the diffusion is also variable.

In order to obtain results of any weight the stars 4 and 3 (κ_1 and κ_2 Tauri), which theoretically should be strongly displaced, must be shown. They appear on all plates from T to Z, and being near the centre of the field have good images. They are relatively rather faint on plate U, but are bright on the other plates. The appearance of the remaining stars is as follows :—

Plate T. 6 bright ; 10 faint.
Plate U. 6, 10 very bright ; 11 faint.
Plate V. 6 bright ; 10 fair.
Plate W. 5, 6 good ; 10 diffused.
Plate X. 5, 6, 11 good.
Plate Y. 5, 6, 11 faint, diffused ; 12 very faint.
Plate Z. 5, 6, 11 faint, diffused.

* It happens that it is also reduced, but to a less extent, by using too small a value of a'.

The possibility of a determination of deflection practically depends on the appearance of star 5. The relative displacement of 5 and 3 is on EINSTEIN's theory, $1''\cdot 2$ in the y-co-ordinate. Further, the x-measures of 5 are needed for a really good determination of the orientation. Star 11 can scarcely take its place. It is true that the relative displacement is then $0''\cdot 8$; but the orientation affects this with a much larger factor, and the orientation is badly determined in the absence of star 5.

Accordingly plates W and X are the only ones likely to give a trustworthy result. X is somewhat the better plate of the two.* Measures have been made of the faint diffused images on plates Y and Z; but, as might have been expected, they are hopelessly discordant and cannot be reconciled by any adopted value of the deflection.

35. We give the measures of plates X and W in detail. Both comparisons of X were measured at Principe a few days after the eclipse. Plate W, which was not developed until after the return of the expedition, was measured at Cambridge on August 22–23.†

Plate X.

(1) Comparison with Oxford Plate G_1.

The differential refraction for all the eclipse plates is

$$a = -46\cdot 5, \qquad b, d = +8\cdot 2, \qquad e = -27\cdot 0$$

the differential aberration being zero.

For the comparison plate G_1

$$a = -19\cdot 1, \qquad b, d = +0\cdot 7, \qquad e = -28\cdot 3.$$

Hence for X – G_1

$$a = -27\cdot 4, \qquad b, d = +7\cdot 5, \qquad e = +1\cdot 3.$$

* Plate X has also the merit of a short exposure, 3s. We should mistrust the x-measures of a long exposure with variable cloud and imperfect guiding, because there is nothing to show that the images of the different stars are formed at the same time.

† Of the comparisons of check plates, $w_1 - b_1$ was measured on August 20, and the others about the end of September. Previous measures had been made at Principe with three earlier check plates taken on the night of May 16; but a slight change of adjustment of tilt was made the following day (thereafter it remained unaltered until the eclipse), and the small change of focus allowed for in the comparisons. These furnished a provisional scale which was used to obtain preliminary results. Afterwards the measurement of check plates was undertaken in a more systematic way, using later plates about which no doubt could arise, and giving the results printed above. No change of any importance was found; the final value for the deflection at the limb was reduced by $0''\cdot 4$ compared with the provisional value, but this was mainly due to the adoption of separate values of a' and e' instead of adopting the mean, and to recalculation of the differential refraction and aberration.

To these must be added the terms representing change of scale, determined from the check plates (Table XIII.), viz.,

$$a = +31\cdot 2, \qquad b, d = -0\cdot 6, \qquad e = +37\cdot 3.$$

Hence the whole difference $X - G_1$ is given by

$$a = +3\cdot 8, \qquad b, d = +6\cdot 9, \qquad e = +38\cdot 6.$$

The first step is to take the measured differences Δx, Δy, and take out the parts $ax + by$, $dx + ey$, due to these terms, leaving the corrected differences $\Delta_1 x$, $\Delta_1 y$.

$\Delta_1 x$ and $\Delta_1 y$ contain (1) the Einstein displacement, if any, and (2) the unknown relative orientation of the plates giving rise to terms of the form, $\Delta x = +\theta y$, $\Delta y = -\theta x$. These two parts could be separated by a least-squares solution, but in view of the poor quality of the material it seems better to adopt a method which keeps a better check on possible discordances and shows more clearly what is happening. The Einstein displacement in x is small, and we might perhaps neglect it altogether in determining θ from the x-measures. However, it is clear from preliminary trials that a displacement exists—whether the half or the full Einstein displacement. Hence if we take out three-quarters of the full Einstein displacement ($\frac{3}{4}E_x$) we divide the already slight effect by 4, and at the same time deal fairly between the two hypotheses.* The residuals $\Delta_2 x$ result.

From the equations $\Delta_2 x = c + \theta y$ we determine by least squares the orientation θ, which is found to be $+163$. Removing the term $163y$ we obtain the residuals $\Delta_3 x$.

Turning to $\Delta_1 y$, we correct for the orientation by taking out the term $-163x$, leaving $\Delta_3 y$. These values should agree for all the stars, except for the displacement and the accidental error.

Denoting the value of the displacement at 50′ (or 10 réseau-intervals) from the centre of the sun by κ, the y-displacements of the various stars will be $\kappa\alpha_y$, where α_y has the values tabulated below. We can therefore obtain κ by solving by least-squares the equations

$$\Delta_3 y = f + \kappa\alpha_y.$$

The radius of the sun during the eclipse was $15'\cdot 78$. Hence the full Einstein displacement of $1''\cdot 75$ corresponds to $0''\cdot 55$ at 50′ distance, or, in our units of $0''\cdot 003$, $\kappa = 184$. It is easily seen that the value is somewhere near this, and it is therefore easier and more instructive to take out $E_y = 184\alpha_y$, and determine the correction to κ from the residuals $\Delta_4 y$. We also remove the mean of $\Delta_4 y$ obtaining the final residuals.

The normal equations corresponding to equations of condition

$$\text{residual} = \delta f + \alpha_y\,\delta\kappa$$

* The smaller the displacement provisionally assumed for x, the larger is the displacement ultimately found from y (see p. 327).

are found to be

$$5\delta f + 2\cdot83\delta\kappa = -1$$

$$2\cdot83\delta f + 4\cdot83\delta\kappa = +64$$

whence

$$3\cdot23\delta\kappa = +64,$$

$$\delta\kappa = +20.$$

An increase of 20 on 184 corresponds to an increase of 0″·19 on 1″·75. Hence the resulting deflection at the limb is 1″·94.

Since the full deflection is indicated we complete the results for x by taking out the remaining $\frac{1}{4}E_x$, obtaining $\Delta_4 x$, and then tabulate the residuals from the mean values −5942.

The successive steps are shown below :—

Star.	x.	Δx.	$3\cdot8x$.	$6\cdot9y$.	$\Delta_1 x$.	$\frac{3}{4}E_x$.	$\Delta_2 x$.	$+163y$.	$\Delta_3 x$.	$\Delta_4 x$.	Resid. (unit = 0″·003).
11	1·39	−3916	5	86	−4007	−76	−3931	2021	−5952	−5927	+ 15
5	12·40	−5518	47	20	−5585	−79	−5506	478	−5984	−5958	− 16
4	17·34	−2869	66	129	−3064	−54	−3010	3051	−6061	−6043	−101
3	17·48	−2924	66	121	−3111	−69	−3042	2869	−5911	−5888	+ 54
6	19·87	−1568	75	172	−1815	+ 3	−1818	4075	−5893	−5894	+ 48

Star.	y.	Δy.	$6\cdot9x$.	$38\cdot6y$.	$\Delta_1 y$.	$-163x$.	$\Delta_3 y$.	E_y.	$\Delta_4 y$.	α_y.	Resid.
11	12·40	6398	10	479	5909	− 227	6136	+ 6	6130	+0·03	+ 5
5	2·93	4121	86	113	3922	−2021	5943	−127	6070	−0·69	− 55
4	18·72	4512	120	722	3670	−2826	6496	+234	6262	+1·27	+137
3	17·60	4236	121	679	3436	−2849	6285	+272	6013	+1·48	−112
6	24·99	4148	137	965	3046	−3239	6285	+136	6149	+0·74	+ 24

(2) Comparison with Oxford Plate H_1.

The reductions are similar and are given in a rather more condensed form below. The theoretical plate constants are

$$a = +3\cdot8, \qquad b, d = +8\cdot3, \qquad e = +38\cdot6.$$

Star.	Δx.	$\Delta_1 x$.	$\Delta_2 x$.	$+10y$.	$\Delta_3 x$.	$\Delta_4 x$.	Resid.
11	7290	7182	7258	124	7134	7159	+235
5	6751	6680	6759	29	6730	6756	−168
4	7126	6905	6959	187	6772	6790	−134
3	7320	7108	7177	176	7001	7024	+100
6	7429	7147	7144	250	6894	6893	− 31

Star.	Δy.	$\Delta_1 y$.	$-10x$.	$\Delta_3 y$.	E_y.	$\Delta_4 y$.	Resid.
11	1586	1095	− 14	1109	+ 6	1103	+172
5	858	642	−124	766	−127	893	− 38
4	1881	1015	−173	1188	+234	954	+ 23
3	1785	961	−175	1136	+272	864	− 67
6	1909	779	−199	978	+136	842	− 89

The normal equations are

$$5\delta f + 2\cdot 83\delta\kappa = +1$$

$$2\cdot 83\delta f + 4\cdot 83\delta\kappa = -105$$

whence

$$3\cdot 23\delta\kappa = -105,$$

$$\delta\kappa = -33.$$

The corresponding deflection at the limb is

$$1''\cdot 75 - 0''\cdot 31 = 1''\cdot 44.$$

Plate W.

Although the exposure was only 10 seconds the images have jumped in R.A., so that the appearance is dumb-bell shaped. They are, however, symmetrical, so that fair measures of x can be made; the y measures on which the result chiefly depends are unaffected. Star 10 is very diffused in R.A.

(1) Comparison with Oxford Plate D_1.

Theoretical plate-constants

$$a = +4\cdot 9, \quad b, d = +6\cdot 5, \quad e = +39\cdot 7.$$

Star.	x.	Δx.	$\Delta_1 x$.	$\frac{3}{4}E_x$.	$\Delta_2 x$.	$+91y$.	$\Delta_3 x$.	$\Delta_4 x$.	Resid.
5	12·40	2450	2370	−79	2449	267	2182	2208	+ 40
4	17·34	3948	3741	−54	3795	1704	2091	2109	− 59
3	17·48	3834	3634	−69	3703	1602	2101	2124	− 44
6	19·87	4525	4266	+ 3	4263	2275	1988	1987	−181
10	22·60	5199	4911	+17	4894	2476	2418	2412	+244

Star.	y.	Δy.	$\Delta_1 y$.	$-91x$.	$\Delta_3 y$.	E_y.	$\Delta_4 y$.	α_y.	Resid.
5	2·93	5320	5123	−1128	6251	−127	6378	−0·69	+ 70
4	18·72	5745	4889	−1578	6467	+234	6233	+1·27	− 75
3	17·60	5911	5098	−1591	6689	+272	6417	+1·48	+109
6	24·99	5628	4507	−1808	6315	+136	6179	+0·74	−129
10	27·21	5616	4389	−2057	6446	+114	6332	+0·62	+ 24

Normal equations

$$5\delta f + 3\cdot42\delta\kappa = -1$$

$$3\cdot42\delta f + 5\cdot21\delta\kappa = -62$$

whence

$$2\cdot87\delta\kappa = -61$$

$$\delta\kappa = -21.$$

Hence deflection at the limb is

$$1''\cdot75 - 0''\cdot20 = 1''\cdot55.$$

(2) Comparison with Oxford Plate I_2.

Theoretical plate constants

$$a = +4\cdot0, \qquad b, d = +9\cdot1, \qquad e = +38\cdot8.$$

Star.	Δx.	$\Delta_1 x$.	$\Delta_2 x$.	$-30y$.	$\Delta_3 x$.	$\Delta_4 x$.	Resid.
5	5050	4973	5052	− 88	5140	5166	+ 46
4	4732	4493	4547	−562	5109	5127	+ 7
3	4622	4392	4461	−528	4989	5012	−108
6	4635	4329	4326	−750	5076	5075	− 45
10	4764	4426	4409	−816	5225	5219	+ 90

Star.	Δy.	$\Delta_1 y$.	$+30x$.	$\Delta_3 y$.	E_y.	$\Delta_4 y$.	Resid.
5	−6824	−7051	372	−7423	−127	−7296	− 15
4	−5751	−6635	520	−7155	+234	−7389	−108
3	−5609	−6451	524	−6975	+272	−7247	+ 34
6	−5425	−6576	596	−7172	+136	−7308	− 27
10	−5109	−6371	678	−7049	+114	−7163	+118

Normal equations

$$5\delta f + 3\cdot42\delta\kappa = +2$$

$$3\cdot42\delta f + 5\cdot21\delta\kappa = -24$$

whence

$$2\cdot87\delta\kappa = -25,$$

$$\delta\kappa = -9.$$

Hence deflection at the limb is

$$1''\cdot75 - 0''\cdot08 = 1''\cdot67.$$

Plate U.

Comparison with Oxford Plate K_2.

Since Plate U shows some good images it has been examined, although owing to the absence of star 8 the weight is small. The measures were made at Principe.

Theoretical plate-constants

$$a = +2\cdot 8, \qquad b, d = +8\cdot 9, \qquad e = +37\cdot 7.$$

Star.	x.	Δx.	$\Delta_1 x$.	$+240y$.	E_x.	$\Delta_4 x$.	Resid.
11	1·39	2905	2791	2976	−101	− 84	−147
4	17·34	4508	4292	4493	− 72	−129	−192
3	17·48	4626	4420	4224	− 92	+288	+225
6	19·87	6270	5992	5998	+ 4	− 10	− 73
10	22·60	7110	6805	6530	+ 23	+252	+189

Star.	x.	Δy.	$\Delta_1 y$.	$-240x$.	E_y.	$\Delta_4 y$.	Resid.
11	12·40	9026	8547	− 334	+ 6	8875	− 94
4	18·72	5846	4986	−4162	+234	8914	− 55
3	17·60	5985	5165	−4195	+272	9089	+120
6	24·99	5458	4339	−4769	+136	8972	+ 3
10	27·21	4911	3684	−5424	+114	8994	+ 25

In this case it is not possible to determine the orientation with sufficient accuracy from the x-measures; the value here applied is an arbitrary preliminary value. We accordingly make a least-squares solution from both x- and y-residuals to determine the correction to the orientation, $\delta\theta$, as well as δc, δf and $\delta\kappa$.

The result is

$$\delta\theta = +2, \qquad \delta\kappa = +121.$$

This gives the deflection

$$2''\cdot 90.$$

The probable error is, however, $\pm\ 0''\cdot 87$, so that the result is practically worthless. Further, it is much more likely to be affected by systematic error than the previous results.

The large probable error is partly due to the large residuals which are greater than in the previous measures; in particular star 3 is unduly faint. If the same accuracy had been obtained, the theoretical weight would have been half that of plates W and X;

but having regard to possible systematic error, probably a quarter weight would more nearly represent the true value.

This determination is ignored in the subsequent discussion.

36. It is easy to calculate the effects of any errors in the adopted scale, orientation, &c., on the final result (deflection at the limb). We give some illustrations.

An error in the adopted scale of y of 10 units (in the fifth place of decimals) would lead to an error $0''\cdot68$ in the result from either plate. Thus the probable error $\pm2\cdot1$ in the determination of e' gives a probable error $\pm0''\cdot14$ in the final result; or, if we adopted the largest (rather discordant) value found for e' instead of the mean, we should reduce the result by $0''\cdot52$.

An error of 10 units in the orientation gives an error in the result of $0''\cdot45$ for plate X, and $0''\cdot22$ for Plate W. It is therefore of less importance, and further it is not likely to be systematic.

Errors in the measurement of x only affect the result through the orientation. For Plate X, a probable error of $\pm0''\cdot20$ in the x-measures would give an error $\pm4\cdot0$ in the orientation, leading to an error $\pm0''\cdot18$ in the result; whereas an error of the same magnitude in the y measures gives directly an error $\pm0''\cdot35$ in the result. For Plate W, the probable error of $\pm0''\cdot20$ in x gives an error $\pm3\cdot5$ in the orientation and $\pm0''\cdot08$ in the result, compared with $\pm0''\cdot38$ for similar inaccuracy in y. It is particularly fortunate that the x-measures are so unimportant for Plate W, because, as already mentioned, the images trailed on that plate.

Finally, it will be remembered that in order not to commit ourselves to the Einstein hypothesis prematurely we neglected the correction $\frac{1}{4}E_x$ in determining the orientation. This will make a difference of $0''\cdot029$ in the results from Plate W and $0''\cdot092$ from Plate X. The effect is that the deduced deflection needs to be decreased, and the mean correction $-0''\cdot06$ should be applied to the mean result obtained, or rather, to make the adopted deflection for x consistent with the deduced value from y, the correction needed is $-0''\cdot04$.

Discussion of the Results.

37. The four determinations from the two eclipse plates are

X − G	$1''\cdot94$
X − H	$1''\cdot44$
W − D	$1''\cdot55$
W − I	$1''\cdot67$

giving a mean of

$1''\cdot65.$

They evidently agree with EINSTEIN's predicted value $1''\cdot75$.

The residuals* in the separate comparisons reduced to arc are as follows. They do not appear to show any special peculiarities.

Star.	x residuals.					y residuals.				
	G.	H.	D.	I.	Mean.	G.	H.	D.	I.	Mean.
	″	″	″	″	″	″	″	″	″	″
11	+0·04	+0·70	—	—	—	+0·01	+0·52	—	—	—
5	−0·05	−0·50	+0·12	+0·14	−0·07	−0·16	−0·11	+0·21	−0·04	−0·02
4	−0·30	−0·40	−0·18	+0·02	−0·21	+0·41	+0·07	−0·22	−0·32	−0·02
3	+0·16	+0·30	−0·13	−0·32	0·00	−0·34	−0·20	+0·33	+0·10	−0·03
6	+0·14	−0·09	−0·54	−0·13	−0·16	+0·07	−0·27	−0·39	−0·08	−0·17
10	—	—	+0·73	+0·27	—	—	—	+0·07	+0·35	—

The average y-residual is $\pm 0''\cdot 22$, which gives a probable error for y of $\pm 0''\cdot 21$. It is satisfactory that this agrees so nearly with the probable error ($\pm 0''\cdot 22$) of the check plates, showing that the images are of about the same degree of difficulty and therefore presumably comparable. The probable error of x is $\pm 0''\cdot 25$, but we are not so much concerned with this.

The weight of the determination of $\delta\kappa$ is about 3 (strictly 3·23 for Plate X and 2·87 for Plate W). The probable error of κ is therefore $\pm 0''\cdot 12$, which corresponds to a probable error of $\pm 0''\cdot 38$ in the final values of the deflection.

As the four determinations involve only two eclipse plates and are not wholly independent, and further small accidental errors may arise through inaccurate determination of the orientation, the probable error of our mean result will be about $\pm 0''\cdot 25$. There is further the error of $\pm 0''\cdot 14$ affecting all four results equally, arising from the determination of scale. Taking this into account, and including the small correction $-0''\cdot 04$ previously mentioned, our result may be written

$$1''\cdot 61 \pm 0''\cdot 30.$$

It will be seen that the error deduced in this way from the residuals is considerably larger than at first seemed likely from the accordance of the four results. Nevertheless the accuracy seems sufficient to give a fairly trustworthy confirmation of EINSTEIN's theory, and to render the half-deflection at least very improbable.

38. It remains to consider the question of systematic error. The results obtained with a similar instrument at Sobral are considered to be largely vitiated by systematic

* The residuals refer to the theoretical deflection $1''\cdot 75$, not the deduced deflections.

errors. What ground then have we—apart from the agreement with the far superior determination with the 4-inch lens at Sobral—for thinking that the present results are more trustworthy ?

At first sight everything is in favour of the Sobral astrographic plates. There are 12 stars shown against 5, and the images though far from perfect are probably superior to the Principe images. The multiplicity of plates is less important, since it is mainly a question of systematic error. Against this must be set the fact that the five stars shown on Plates W and X include all the most essential stars; stars 3 and 5 give the extreme range of deflection, and there is no great gain in including extra stars which play a passive part. Further, the gain of nearly two extra magnitudes at Sobral must have meant over-exposure for the brighter stars, which happen to be the really important ones; and this would tend to accentuate systematic errors, whilst rendering the defects of the images less easily recognised by the measurer. Perhaps, therefore, the cloud was not so unkind to us after all.

Another important difference is made by the use of the extraneous determination of scale for the Principe reductions. Granting its validity, it reduces very considerably both accidental and systematic errors. The weight of the determination from the five stars with known scale is more than 50 per cent. greater than the weight from the 12 stars with unknown scale. Its effect as regards systematic error may be seen as follows. Knowing the scale, the greatest relative deflection to be measured amounts to $1''\cdot 2$ on EINSTEIN'S theory; but if the scale is unknown and must be eliminated, this is reduced to $0\cdot''67$. As we wish to distinguish between the full deflection and the half deflection, we must take half these quantities. Evidently with poor images it is much more hopeful to look for a difference of $0''\cdot 6$ than for $0''\cdot 3$. It is, of course, impossible to assign any precise limit to the possible systematic error in interpretation of the images by the measurer; but we feel fairly confident that the former figure is well outside possibility.

A check against systematic error in our discussion is provided by the check plates, as already shown. Its efficacy depends on the similarity of the images on the check plates and eclipse plates at Principe. Both sets are fainter than the Oxford images with which they are compared, the former owing to the imperfect driving of the cœlostat, which made it impossible to secure longer exposures, the latter owing to cloud. Both sets have a faint wing in declination, but this is separated by a slight gap from the true images, and, at least on the plates measured, the wing can be distinguished and ignored. The images on Plates W and X are not unduly diffused except for No. 10 on Plate W. Difference in quality between the eclipse images and the Principe check images is not noticeable, and is certainly far less than the difference between the latter and the Oxford images; and, seeing that the latter comparison gives no systematic error in y, it seems fair to assume that the comparison of the eclipse plates is free from systematic error.

The writer must confess to a change of view with regard to the desirability of using an extraneous determination of scale. In considering the programme it had seemed too risky a proceeding, and it was thought that a self-contained determination would receive more confidence. But this opinion has been modified by the very special circumstances at Principe; and it is now difficult to see that any valid objection can be brought against the use of the scale.

The temperature at Principe was remarkably uniform and the extreme range probably did not exceed 4° during our visit—including day and night, warm season and cold season. The temperature ranged generally from $77\frac{1}{2}°$ to $79\frac{1}{2}°$ in the rainy season, and about 1° colder in the cool gravana. All the check plates and eclipse plates were taken within a degree of the same temperature, and there was, of course, no perceptible fall of temperature preceding totality. To avoid any alteration of scale in the daytime the telescope tube and object-glass were shaded from direct solar radiation by a canvas screen; but even this was scarcely necessary, for the clouds before totality provided a still more efficient screen, and the feeble rays which penetrated could not have done any mischief. A heating of the mirror by the sun's rays could scarcely have produced a true alteration of scale though it might have done harm by altering the definition; the cloud protected us from any trouble of this kind. At the Oxford end of the comparison the scale is evidently the same for both sets of plates, since they were both taken at night and intermingled as regards date.

It thus appears that the check scale is legitimately applicable to the eclipse plates. But the method may not be so satisfactory at future eclipses, since the particular circumstances at Principe are not likely to be reproduced. As regards other sources of systematic error, our chief guarantee lies in the comparatively large amount of the deflection to be measured, and the test satisfied by the check plates that photographs of another field under similar conditions show no deflections comparable with those here found.

V. General Conclusions.

39. In summarising the results of the two expeditions, the greatest weight must be attached to those obtained with the 4-inch lens at Sobral. From the superiority of the images and the larger scale of the photographs it was recognised that these would prove to be much the most trustworthy. Further, the agreement of the results derived independently from the right ascensions and declinations, and the accordance of the residuals of the individual stars (p. 308) provides a more satisfactory check on the results than was possible for the other instruments.

These plates gave

From declinations	$1''\cdot94$
From right ascensions	$2''\cdot06$

The result from declinations is about twice the weight of that from right ascensions, so that the mean result is

$$1''\cdot 98$$

with a probable error of about $\pm 0''\cdot 12$.

The Principe observations were generally interfered with by cloud. The unfavourable circumstances were perhaps partly compensated by the advantage of the extremely uniform temperature of the island. The deflection obtained was

$$1''\cdot 61.$$

The probable error is about $\pm 0''\cdot 30$, so that the result has much less weight than the preceding.

Both of these point to the full deflection $1''\cdot 75$ of EINSTEIN'S generalised relativity theory, the Sobral results definitely, and the Principe results perhaps with some uncertainty. There remain the Sobral astrographic plates which gave the deflection

$$0''\cdot 93$$

discordant by an amount much beyond the limits of its accidental error. For the reasons already described at length not much weight is attached to this determination.

It has been assumed that the displacement is inversely proportional to the distance from the sun's centre, since all theories agree on this, and indeed it seems clear from considerations of dimensions that a displacement, if due to gravitation, must follow this law. From the results with the 4-inch lens, some kind of test of the law is possible though it is necessarily only rough. The evidence is summarised in the following table and diagram, which show the radial displacement of the individual stars (mean from all the plates) plotted against the reciprocal of the distance from the centre. The displacement according to EINSTEIN'S theory is indicated by the heavy line, according to the Newtonian law by the dotted line, and from these observations by the thin line.

RADIAL Displacement of Individual Stars.

Star.	Calculation.	Observation.
	"	"
11	0·32	0·20
10	0·33	0·32
6	0·40	0·56
5	0·53	0·54
4	0·75	0·84
2	0·85	0·97
3	0·88	1·02

Thus the results of the expeditions to Sobral and Principe can leave little doubt that a deflection of light takes place in the neighbourhood of the sun and that it is of the amount demanded by EINSTEIN'S generalised theory of relativity, as attributable to the sun's gravitational field. But the observation is of such interest that it will probably be considered desirable to repeat it at future eclipses. The unusually favourable conditions of the 1919 eclipse will not recur, and it will be necessary to photograph fainter stars, and these will probably be at a greater distance from the sun.

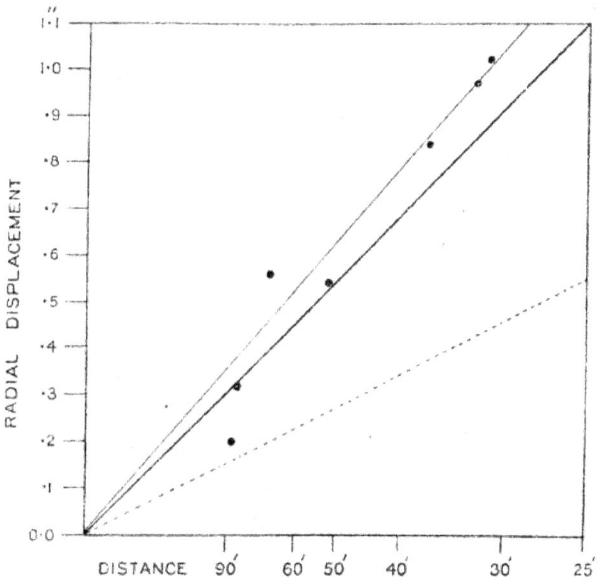

Diagram 2.

This *can* be done with such telescopes as the astrographic with the object-glass stopped down to 8 inches, if photographs of the same high quality are obtained as in regular stellar work. It will probably be best to discard the use of cœlostat mirrors. These are of great convenience for photographs of the corona and spectroscopic observations, but for work of precision of the high order required, it is undesirable to introduce complications, which can be avoided, into the optical train. It would seem that some form of equatorial mounting (such as that employed in the Eclipse Expeditions of the Lick Observatory) is desirable.

In conclusion, it is a pleasure to record the great assistance given to the Expeditions from many quarters. Reference has been made in the course of the paper to some of these. Especial thanks are due to the Brazilian Government for the hospitality and facilities accorded to the observers in Sobral. They were made guests of the

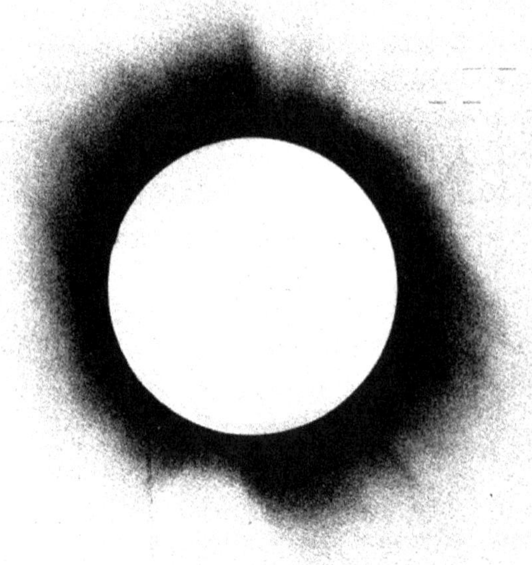

Government, who provided them with transport, accommodation and labour. Dr. MORIZE, Director of the Rio Observatory, acting on behalf of the Brazilian Government, made most complete arrangements for the Expedition, and in this way contributed materially to its success.

On behalf of the Principe Expedition, special thanks are due to Sr. JERONYMO CARNEIRO, who most hospitably entertained the observers and provided for all their requirements, and to Sr. ATALAYA, whose help and friendship were of the greatest service to the observers in their isolated station.

We gratefully acknowledge the loan for more than six months of the astrographic object-glass of the Oxford University Observatory. We are also indebted to Mr. BELLAMY for the check plates he obtained in January and February.

Thanks are due to the Royal Irish Academy for the loan of the 4-inch object-glass and 8-inch cœlostat.

As stated above, the expeditions were arranged by the Joint Permanent Eclipse Committee with funds allocated by the Government Grant Committee.

[In Plate 1 is given a half-tone reproduction of one of the negatives taken with the 4-inch lens at Sobral. This shows the position of the stars, and, as far as possible in a reproduction of this kind, the character of the images, as there has been no retouching.

A number of photographic prints have been made and applications for these from astronomers, who wish to assure themselves of the quality of the photographs, will be considered and as far as possible acceded to.]

www.ingramcontent.com/pod-product-compliance
Lightning Source LLC
Chambersburg PA
CBHW062006200326
41519CB00017B/4690
9781927763285